Lecture Notes in Statistics

Edited by J. Berger, S. Fienberg, J. Gani, K. Krickeberg, and B. Singer

56

James K. Lindsey

The Analysis of Categorical Data Using GLIM

Springer-Verlag
New York Berlin Heidelberg London Paris Tokyo Hong Kong

Author

James K. Lindsey
Faculté d'Economie, de Gestion et de Sciences Sociales
Université de Liège, 4000 Liège, Belgium

Mathematical Subject Classification Codes: 62-02, 62-04, 62-07, 62H17, 62P10, 62P25

ISBN 0-387-97029-0 Springer-Verlag New York Berlin Heidelberg
ISBN 3-540-97029-0 Springer-Verlag Berlin Heidelberg New York

Printed in Germany

Printing and binding: Druckhaus Beltz, Hemsbach/Bergstr.
2847/3140-543210 – Printed on acid-free paper

Preface

The present text is the result of teaching a third year statistical course to undergraduate social science students. Besides their previous statistics courses, these students have had an introductory course in computer programming (FORTRAN, Pascal, or C) and courses in calculus and linear algebra, so that they may not be typical students of sociology. This course on the analysis of contingency tables has been given with all students in front of computer terminals, and, more recently, microcomputers, working interactively with GLIM.

Given the importance of the analysis of categorical data using log linear models within the overall body of models known as general linear models (GLMs) treated by GLIM, this book should be of interest to anyone, in any field, concerned with such applications. It should be suitable as a manual for applied statistics courses covering this subject.

I assume that the reader has already a reasonably strong foundation in statistics, and specifically in dealing with the log-linear/logistic models. I also assume that he or she has access to the GLIM manual and to an operational version of GLIM itself. In other words, this book does not pretend to present either a complete introduction to the use of GLIM or an exposition of the statistical properties of log-linear/logistic models. For the former, I would recommend Healy (1988) and Aitkin et al (1989). For the latter, many books already exist, of which I would especially recommend that of Fingleton (1984) in the present context.

In this book, I attempt to show how the GLIM statistical system can easily be applied to a wide variety of log-linear/logistic models. i.e. the interface between statistical analysis and computer use. To get the most out of the book, it is important to try out the examples with GLIM as one proceeds.

All of the present book, including the statistical analysis of all of the examples, using GLIM 3.77 update 2, and the word processing and page setting, using ProWrite and ProScript, was produced by the author on a 2.5 megabyte Commodore Amiga microcomputer.

Many of the GLIM macros in Appendix III were originally written by various authors, cited therein. They have been modified to make them user-friendly and, hence, more accessible to someone not familiar with the details of GLIM programming. May the original authors be thanked for making their work public.

I would like to thank my students over the past ten years who have suffered through this course and supplied invaluable reactions and comments and R. Doutrelepont who supplied the data for the example on the Belgian elections in Chapter 6.

TABLE OF CONTENTS

CHAPTER 1

ONE-WAY FREQUENCY TABLES

1. A Time Trend Model

The simplest frequency tables concern a single variable and show the frequencies with which the various categories of that variable have been observed. Here, we shall be interested in frequency tables where the variable may be nominal, discrete, or continuous, but where the only assumption is that of a multinomial distribution. In those cases where the categories refer to some continuous measure, such as income, length of employment, etc. or are themselves counts, such as numbers of accidents per individual, number of children per family, etc., specific probability distributions can often be fitted to the data. However, this latter problem is not the subject of the present book.

Having stated this restriction, we must immediately qualify it, since our tool of analysis, GLIM, does not directly handle the multinomial distribution. Nevertheless, we can very simply demonstrate that models based on the multinomial distribution

$$\binom{F.}{F_1 \ldots F_K} p_1^{F_1} \ldots p_K^{F_K} \tag{1.1}$$

where $F_1 \ldots F_K$ are the frequencies and $p_1 \ldots p_K$ the corresponding probabilities, can equivalently be analysed by models which GLIM does treat, those based on the Poisson distribution

$$\frac{e^{-\mu} \mu^{F}}{F!} \tag{1.2}$$

for each category, if we condition on the total number of observations. Let us recall two points of probability theory. First, a conditional distribution is defined by

$$\Pr(A/B) = \Pr(A \text{ and } B)/\Pr(B) \tag{1.3}$$

Second, if a set of frequencies, $F_1 \ldots F_K$, have a Poisson distribution with means $\mu_1 \ldots \mu_K$, then their sum F. also has a Poisson distribution with mean $\mu_.$, the sum of the individual means.

We are now in a position to demonstrate the relationship between the multinomial and the conditional Poisson distributions:

$$\prod \frac{e^{-\mu_k} \mu_k^{F_k}}{F_k!} \Big/ \frac{e^{-\mu_.} \mu_.^{F.}}{F.!} = \frac{F.!\, e^{-\mu_.} \prod \mu_k^{F_k}}{\prod F_k!\, e^{-\mu_.} \mu_.^{F.}}$$

$$= \binom{F_{.}}{F_1 \ldots F_K} \Pi \left(\frac{\mu_k}{\mu_{.}}\right)^{F_k}$$

so that $p_k = \mu_k/\mu_{.}$ and the two distributions are identical.

Before going further, we shall consider our first example (Table 1.1). The tables for all examples, in a form ready for GLIM, along with the GLIM instructions to produce the output shown in the text, are also provided in Appendix II.

Months Before	1	2	3	4	5	6	7	8	9	10	11	12	13	14	15	16	17	18
Number	15	11	14	17	5	11	10	4	8	10	7	9	11	3	6	1	1	4

Table 1.1 Subjects Reporting One Stressful Event (Haberman, 1978, p.3)

Our variable is the number of months prior to an interview that subjects remember a stressful event. We wish to determine if the probability is the same for remembering such an event in all of these 18 months. If we look at Table 1.1, we see immediately that the number of events remembered seems, in fact, to decrease with time.

GLIM is an interactive system which reads and interprets each instruction as it receives it. As soon as it has obtained sufficient information, it executes the instruction. A question mark (?) usually indicates that it is ready for further input. A log or transcript of the entire session is normally produced on a file for the user.

To start any analysis, GLIM requires certain basic information. We must

(1) define the standard vector length - $UNits n

(This refers to the variables to be analysed.)

(2) provide the list of names of variables or vectors into which the data are to be read - $DAta name list

(3) read the values to be placed in the vectors with these names- $Read data list

(4) specify which is the dependent variable - $Yvariate variable name

(5) specify the probability distribution - $ERror distribution

(6) fit the desired linear model - $Fit model

From this first list of six instructions, we notice that all GLIM instructions begin with a dollar sign ($). We may place as many instructions as we like on a line or one on each line. However, an important point, names in GLIM, including instructions and variables, have only four significant characters. Thus, it is sufficient to type $UNI, $DAT, $REA, $YVA, $ERR, and $FIT. All subsequent characters are ignored until a blank or another $ is encountered. Many instructions may be even further shortened as can be seen in Appendix I. Throughout the text, the shortest allowable form will be indicated by capital letters. Variable names must begin with a letter and, of course, cannot contain a $, or other punctuation or operators. They can contain numbers. Special care must be taken with the four character maximum since, for example, GLIM does not distinguish between the variable names CLASS1 and CLASS2.

A model is specified in $Fit by the list of variables which it contains, each separated by an operator (+, -, ., *, /). The + may be used to add a variable to the previous model fitted and - to remove one. The . and * signify interactions (to be explained below). The / will not be used in this text.

We can now construct our first program:

```
$UNits 18 $DAta FREQ $Read 15 11 14 17 5 11 10 4 8 10 7 9 11 3 6 1 1 4
$Yvariate FREQ $ERror P $Fit $
```

Most of this should be clear from what preceded. $ERror P specifies a Poisson distribution. $Fit followed by nothing but $ specifies the fit of a general mean. The output from our program is as follows:

```
    scaled deviance = 50.843 at cycle  4
                d.f. = 17
```

The Poisson distribution in such models is fitted by successive approximations. Here, four iterations were required as indicated by the cycle. For such models, the scaled deviance yields a Chi-square with the degrees of freedom (d.f.) shown. If we look up this value in a table, we see that a Chi-square of 50.84 with 17 d.f. reveals a very significant lack of fit, indicating that the probability of recall is not the same for all of the months.

With one additional instruction, we may go further in our study of this model applied to these data:

(7) provide further information about the fitted model - $Display code for information desired

Then, after the $Fit of our program, we enter

```
$Display ER
```

This signifies that we desire the parameter estimates (E) and the fitted values and residuals (R). The output is:

```
         estimate          s.e.      parameter
      1        2.100      0.08248         1
      scale parameter taken as   1.000

    unit   observed      fitted     residual
      1          15      8.167        2.391
      2          11      8.167        0.991
      3          14      8.167        2.041
      4          17      8.167        3.091
      5           5      8.167       -1.108
      6          11      8.167        0.991
      7          10      8.167        0.642
```

8	4	8.167	-1.458
9	8	8.167	-0.058
10	10	8.167	0.642
11	7	8.167	-0.408
12	9	8.167	0.292
13	11	8.167	0.991
14	3	8.167	-1.808
15	6	8.167	-0.758
16	1	8.167	-2.508
17	1	8.167	-2.508
18	4	8.167	-1.458

We must now elaborate on the model which we have fitted. This is what is commonly called a log linear model, since it is linear in the logarithms of the frequencies. Specifically, we have fitted a common mean to all of the frequencies:

$$\log(F_k) = \mu \qquad \text{for all } k \tag{1.4}$$

The maximum likelihood estimate is $\mu = 2.100$ with standard error 0.08248. As a rough indicator, if the absolute value of a parameter estimate is at least twice the standard error, the estimate is significantly different from zero at the 5% level. This has little relevance in the present case, but is very useful when a large number of parameters are present in the model, since we then have a quick indication of which variables might be eliminated.

We next note that all observations are estimated by the same fitted value, 8.167, since our model only contains the mean. The residuals are differences between observed and fitted values standardized by their standard errors.

We should note that all of our analysis up until now applies to any set of frequencies whether structured or not. Our variable could have been nominal since we have not yet used the ordering of the months.

Let us now examine the residuals more closely. We see that the first four are positive and the last five are negative, indicating that the probability of recalling an event is more than average in the recent months and less than average in the longer time period.

We may now introduce this ordering so as to study the observed decrease in number of events remembered. Suppose that the probability of remembering an event diminishes in the same proportion between any two consecutive months:

$$p_k/p_{K-1} = \emptyset \quad \text{(a constant for all k)} \tag{1.5}$$

Then

$$p_k/p_1 = \emptyset^{k-1} \tag{1.6}$$

and

$$\log(p_k/p_1) = (k-1)\log(\emptyset)$$

but

$$p_k = F_k/F.$$

so that

$$\log(p_k/p_1) = \log(F_k/F_1)$$

and

$$\log(F_k) = \log(F_1) + (k-1)\log(\emptyset)$$
$$= \log(F_1/\emptyset) + k\log(\emptyset)$$

which may be rewritten

$$\log(F_k) = \beta_0 + \beta_1 k \qquad (1.7)$$

where

$$\beta_0 = \log(F_1/\emptyset) \text{ and } \beta_1 = \log(\emptyset)$$

This is a log linear time trend model, a special case of linear regression. To perform the GLIM analysis for this model, we must construct a variable for months. Any such arithmetic calculation may be performed in the following way:

(8) perform an arithmetic calculation - $CAlculate arithmetic expression

The standard arithmetic operators (+, -, *, /) may be used. Note, however, that they all have a different meaning than in $Fit. Note, also, that $CAlculate in GLIM performs vector operations automatically in a way similar to the programming language APL. We, then, enter the following instructions:

```
$CAlculate MON=%GL(18,1) $Fit MON $Display ER
```

In our calculation, we have used one of the GLIM functional operators, %GL(k,n), which fills a vector with integers from 1 to k in blocks of n. For example, with a vector of length 5, %GL(3,2) constructs the vector (1, 1, 2, 2, 3). In our case, we obtain a vector of length 18 filled with the integers from 1 to 18, as required. We, then, fit the model and display the required information, as in the previous case. The output is:

```
scaled deviance = 24.570 at cycle  4
         d.f. = 16

       estimate          s.e.        parameter
  1       2.803        0.1482        1
  2    -0.08377       0.01680        MON
  scale parameter taken as   1.000

unit   observed     fitted    residual
  1        15       15.171      -0.044
  2        11       13.952      -0.790
  3        14       12.831       0.326
  4        17       11.800       1.514
  5         5       10.852      -1.776
  6        11        9.980       0.323
  7        10        9.178       0.271
  8         4        8.440      -1.528
  9         8        7.762       0.085
 10        10        7.138       1.071
 11         7        6.565       0.170
 12         9        6.037       1.206
 13        11        5.552       2.312
 14         3        5.106      -0.932
 15         6        4.696       0.602
 16         1        4.318      -1.597
 17         1        3.971      -1.491
 18         4        3.652       0.182
```

By adding one parameter, and losing one degree of freedom, we have reduced our deviance by 50.84-24.57=26.27. This is a Chi-square with 17-16=1 d.f. which is highly significant, indicating that we must reject our equal probability model in favour of the one for constant reduction in probability. The remaining deviance of 24.57 is a Chi-square with 16 d.f., which is not significant at 5%, indicating that non-linear effects, a non-constant reduction in probability, need not be taken into account.

It would be useful to have the Chi-square probability level directly on the screen instead of referring each time to a table. This is possible by means of a macro or small prefabricated program, supplied in Appendix III, written in the GLIM programming language. This is a language very similar in syntax to the C language, but is interactive.

To load and use such a macro, we need three new GLIM instructions:

(9) read one or more programs or program segments stored on file - $INput file number and program name(s)

(10) specify information to be used in a macro - $Argument macroname and parameter list (max. 9 items)

(11) execute a macro (program) - $Use macroname

Instructions (10) and (11) may be combined by placing the parameter list after the macroname in $Use.

To use the macro, the Chi-square value (the deviance) and the degrees of freedom must be supplied. After each fit, these are available in the scalars %DV and %DF respectively. At each new fit, the old values are lost if they are not stored in other scalars. At our present stage of analysis, we shall have %DV=24.57 and %DF=16. We enter the following instructions:

```
$CAlculate %A=50.84-%DV $CAlculate %B=17-%DF $INput 12 CHIT
$Use CHIT %DV %DF $Use CHIT %A %B
```

The output is

```
Chi2 probability =  0.0775 for Chi2 =   24.57 with   16. d.f.

Chi2 probability =  0.0000 for Chi2 =   26.27 with   1. d.f.
```

confirming the above results.

We note, in the above instructions, that the macro is called CHIT and is found on a file referred to by the number 12. When this instruction is typed, an explanation of the use of the macro appears on the screen.

In applying this macro, we have introduced the use of scalars. Two types exist in GLIM. System scalars consist of the % with two or three letters and contain values produced by certain instructions. (Note, however, that system vectors are also represented by % with two letters.) They are listed in Appendix I. Ordinary scalars are % with one letter and are all initialised to zero. The user manipulates them with $CAlculate.

Our estimated model is

$$\log (F_k) = 2.803 - 0.08377\, k$$

The negative value of β_1, the slope parameter for months, indicates decrease in probability with time elapsed. Since $\beta_1 = \log(\phi)$, $\phi = e^{\beta_1}$ so that $\phi = 0.9196$, the proportional decline in probability per month. If we rewrite our model (1.5) in terms of β_1, we have

$$p_k = p_1 \phi^{k-1} = p_1 e^{\beta_1 (k-1)} \qquad (1.8)$$

which is a model of exponential decay. If β_1 were positive, it would be a model of exponential growth.

It is now possible to plot our model with GLIM. The instruction is

(12) plot several variables on a scattergram - $Plot ordinates and abscissa

The fitted values for the model are contained in the system vector called %FV. We plot observed and fitted values against the month using

```
$Plot %FV FREQ MON
```

where %FV and FREQ are variables for the ordinate and MON for the abscissa. This gives

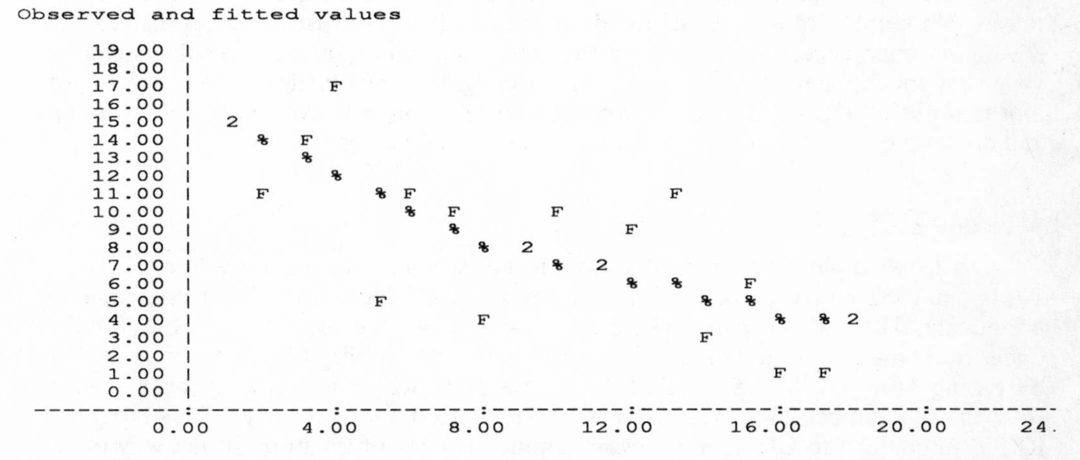

The observed values are represented by F (for FREQ) and the fitted values by % (for %FV). When two or more points fall at the same place, they are represented by a number between 2 and 9 instead of by the first character of the vector. The user also has the option of choosing any other symbol to represent each vector.

We see the form of the exponential decay in the curved line of %s. If we take logarithms of the observed and fitted values using $CAlculate and the function %LOG, and plot them,

```
$CAlculate F=%LOG(FREQ) $CAlculate T=%LOG(%FV) $Plot T F MON
```

we obtain

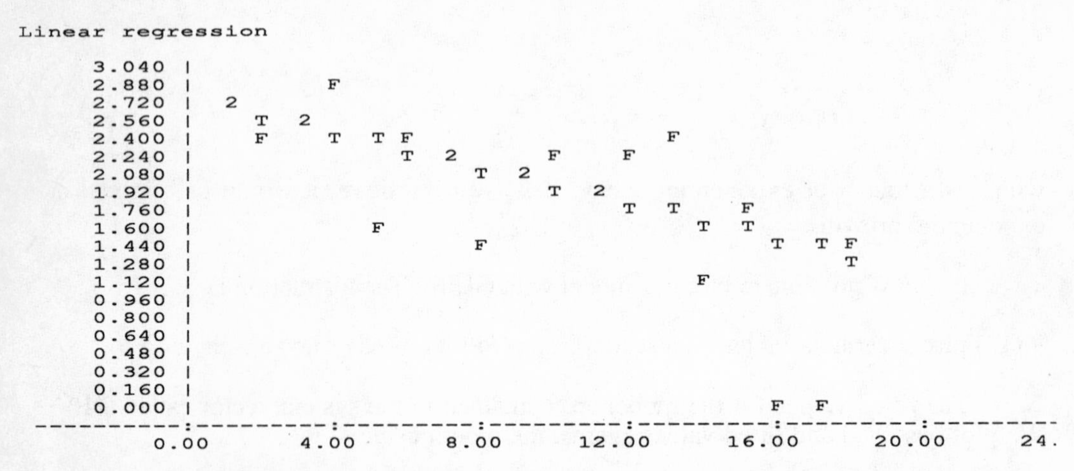

which is our linear model represented by the straight line of Ts (the slight wobble occurs because characters cannot be printed in between lines), surrounded by the observed points, F.

We have seen the usefulness of interpreting the residuals of a model. We should note, however, that inspection of residuals only proves useful when we have a reasonable number of degrees of freedom. As the degrees of freedom approach zero, the model must necessarily represent the data more closely and the residuals cannot vary very much from zero. The task of inspecting residuals is also made easier if we plot them using GLIM. This time, we use a macro (found in Appendix III) to set up and do the plotting. The additional instructions to be typed are

```
$INput 23 RESP $Use RESP
```

A residual plot and a plot of a score test coefficient of sensitivity (see Gilchrist, 1981 and 1982 and Pregibon, 1982) are provided for the model fitted immediately previously. Thus, at the point where we now are, we can only obtain plots for the linear trend model. To obtain those for the constant probability model, we must refit it by typing $Fit and then $Use RESP. The program RESP has already been loaded from the file number 23, so that $INput 23 RESP need not be repeated. This program, RESP, in contrast to CHIT, requires no supplementary information, in the way of an argument list, for its use.

The plots for the two models are as follows:

(1) constant probability model

Poisson Residuals

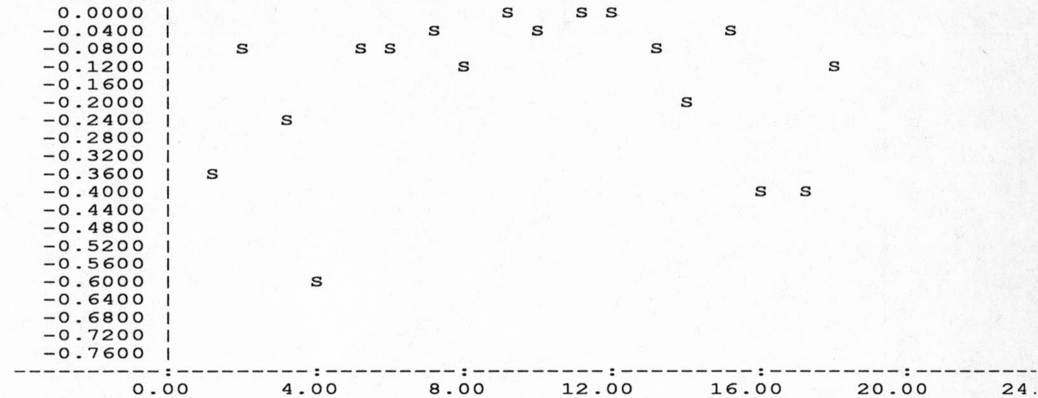

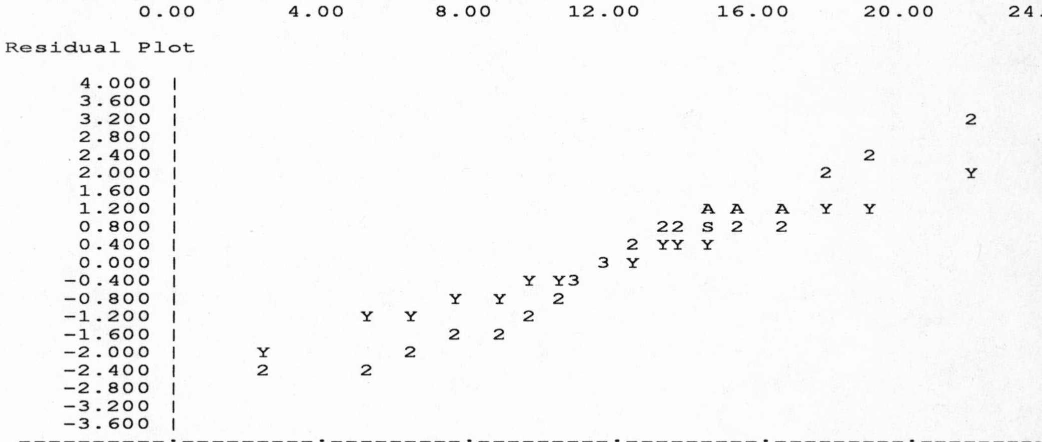

Points Y represent 45 line

(2) time trend model

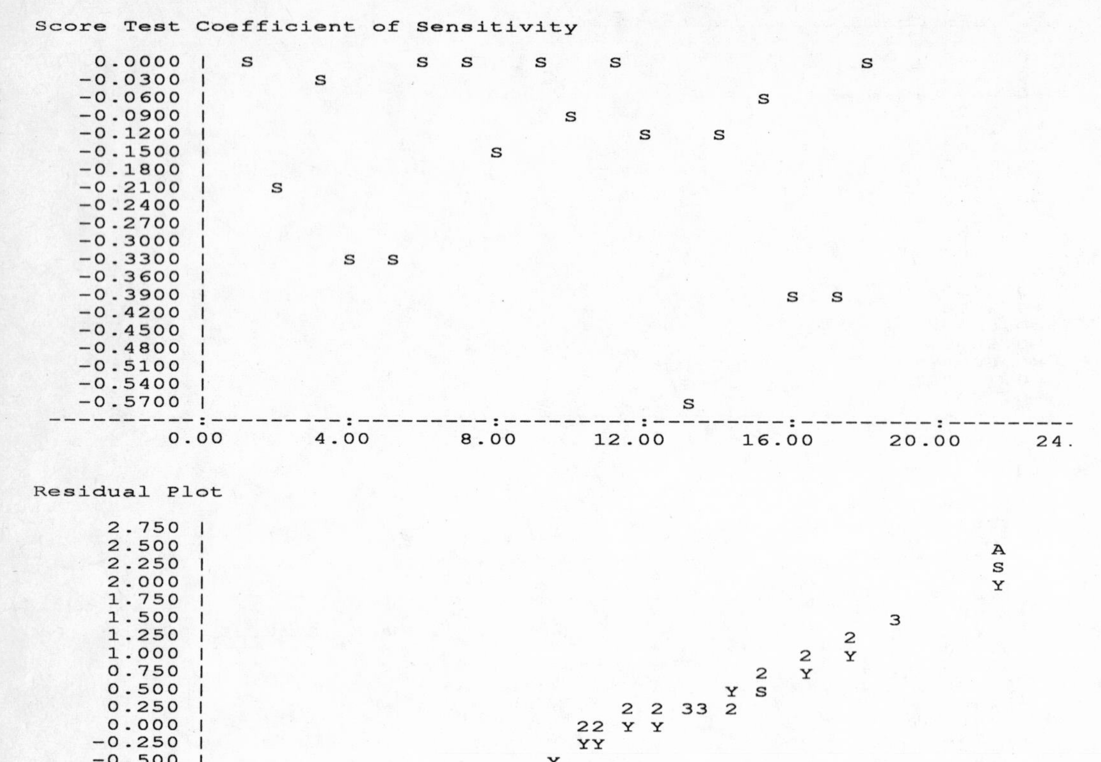

If a model is acceptable, the residual plot should be on a 45 degree straight line, coinciding with that represented by the Y points on the graph. We observe that the residual plot for the constant probability model has a slope greater than this, indicating the lack of fit whereas that for the log linear trend model has the required slope. The score test coefficient is a plot of the individual observations in order as against a modified residual. The model fits least well those observations with large negative values of this coefficient. For the constant probability model, we see that the first and last observations fit less well as we already noticed above. For the time trend model, no obvious pattern is observable, although the thirteenth month fits least well. From the list of residuals given above, we see that 11 recalls are recorded while only 5.6 are predicted.

Finally, to terminate the session and exit from GLIM, we use the instruction:

(13) end the session - $STop

2. Further GLIM Instructions

Before introducing the next model, we shall consider some further details about GLIM which facilitate programming or improve presentation of results. Only the minimum details for directives are presented here; for a full presentation, see the GLIM manual (Payne, 1985).

A number of special symbols may be used. The $ alone means to execute immediately the preceding instruction; this is especially necessary when GLIM does not know if the list of information after an instruction is complete or not. This is indicated by GLIM reproducing the previous instruction on the screen, instead of simply a ?. Such is the case, for example, with $CAlculate and $Fit, since the arithmetic expression or model definition may extend over several lines and execution may not begin before it is complete. In contrast, $Display may execute for each symbol (E, R, etc.) independently, so that the terminating $ is not required. In the same way, $INput may load each program separately.

The symbol : indicates repetition of the immediately preceding command, perhaps with different information following it. For example,

```
$CAlculate PW=1 : X=2
```

The symbol # attached to the name of a macro has the same effect as $Use when the macro contains a series of instructions. For example,

$Use CHIT and #CHIT have the same effect.

However, an argument list cannot follow when # is used in the way it can with $Use.

The symbol, !, the end of line character, causes GLIM to ignore everything which follows on the line. Its use can make programs much more efficient, since otherwise GLIM must read and check (interpret) every character on every line, even if most of it is blank, just to verify if there is something further along. Placing this symbol, !, after the last instruction on the line eliminates this verification.

If an operator appears to the left of the leftmost equals sign in $CAlculate, the answer will be printed out. For example, $CAlculate 3+4 prints 7. Note that $CAlculate 0+I=J+K will print out a column of all values assigned to the vector I.

The instruction $INput may be used to load GLIM instructions which the user has previously placed in a file. In the context of this book, this file will be number 5. The instructions may be placed directly in the file in the same way as if typed interactively on the screen, but the last line of the file must contain, starting in the first column, the instruction:

(14) indicate the end of a program file - $FINish

If the instructions loaded by $INput 5 are not given a macro name, they are all read and immediately executed, as if typed to the screen from the keyboard. But they are not stored and cannot be reused in the way a macro is. In such files, the ! symbol may be used to indicate end of line so that descriptive information may be included in the file, as in the comments to the programs and macros in Appendices II and III.

Often data are too voluminous to be simply typed after $Read or they already exist on a file (which must contain only numerical values, at least in the columns to be read by GLIM). These may be read by

(15) read numerical data from a file - $DINput file number

Here, by convention in this book, the file number will be 1. As with $Read, $DINput must be preceded by $DAta with the list of variable names. If more than one variable name is listed, the values must be presented individual by individual, i.e. the first value in order for each variable, then the second and so on. In this way, the data are read in free format; they must have one or more spaces between each value. In such files, the ! symbol may be used to indicate end of line so that descriptive information may be included in the file, as in the tables of Appendix II.

If the data occupy consecutive columns of a file and, thus, do not have the necessary spaces, as is often the case for large data sets, a format, defining which columns are to be read, must be specified before $Read or $DINput:

(16) format specification - $FOrmat FORTRAN format in parentheses

The FORTRAN format specifies the columns where the values are found (F) and the columns to be ignored (X). For example, three variables, SEX, REVENUE, AGE, to be read in columns 3-6, 7 and 13-14, and other data not presently needed in the remaining columns, require the instructions

```
$DAta SEX REV AGE
$FOrmat (2X,F4.0,F1.0,5X,F2.0)
$DINput 1
```

Any columns not specified by F are ignored, even if they contain values. The ! symbol may not be used in a data file to be read with $FOrmat unless it always appears in columns which are not read.

The output from GLIM instructions can be written on a file rather than appearing on the screen with the instruction:

(17) write results on a file - $OUtput file number

In the context of this book, the file number 6 will be used. After $OUtput 6, all results are written to the file until $OUtput 9 (the number for the terminal on the Commodore Amiga) is typed, at which time they begin again on the screen. The file produced will be clean, in that the GLIM instructions are not mixed with the results. The programs in Appendix II, used to produce the output throughout the text, are presented in this way. However, use of this system may sometimes mean that one does not know to what the results refer. The accompanying GLIM instructions may be printed on the file by the instruction

(18) print out all information received by GLIM - $ECho

before starting the output. The process is stopped by giving the same instruction a second time. This method will produce a file similar, in many respects, to the transcript file normally resulting from a GLIM session.

Titles and other indications may be written with the instructions

(19) write a message - $PRint information

(20) list values of scalars or vectors - $LOok scalar or vector names

After $PRint, text must be included in quotes and must not contain a $. For example, $PRint 'This is the title'. The symbol / after $PRint will cause a new page to begin, if the file is printed on a line printer. Vectors and scalars may also be printed by listing their names after the instruction. These may be mixed with text. Vectors are printed in lines. In contrast, $LOok lists the values of the vectors in columns. Finally, $PRint may be used to print the contents of a macro, whether to the screen or to a file. The actual contents of the macro are printed, whether instructions or text, so that this method may be used to print repeatedly the same title, as in all examples of Appendix II.

We have already mentioned several times the use of macros. An important application is to retain in memory a series of instructions which are to be used repeatedly instead of retyping them each time. The form of a macro is

(21) define a macro or program segment - $Macro macroname space text $End

If the text is a series of instructions, they are executed by typing $Use macroname or #macroname. Macros follow the same rules for naming as do vector variables, and compete for space in memory with them.

System information, such as memory space available, may be obtained by

(22) present system information - $ENVironment code for information desired

The code D yields a list of vectors and macros defined with the space used by each, while U gives space available, S a list of the system structures, and I the special characteristics of your implementation of GLIM.

If GLIM gives an error message that too many structures are defined for the memory, the unnecessary ones may be removed by

(23) delete unwanted user-defined structures - $DElete list of macronames and/or variable names

The list of all such structures present may, of course, first be obtained by $ENVironment D.

Certain system vectors, such as the parameter estimates, are not directly available, but may only be obtained by the instruction

(24) supply a system structure - $EXTract structure

The following two instructions permit the construction of contingency tables from raw observations and their visualization. They will not be used in what follows, since the tables are already created.

(25) create a table from raw data - $Tabulate For variable list separated by semi-colons Using name of table to be created By new variables to be created separated by semi-colons

After appropriately redefining $UNits and specifying $Yvariate as the table created, one may proceed to fit models containing the new variables.

(26) display a table - $TPrint vector containing frequencies and list of variables classifying table separated by semi-colons

Finally, to terminate analysis of a data set without leaving GLIM

(27) end a data analysis - $End

This instruction deletes all vectors and macros previously defined in preparation for analysis of a new data set. Note that, because GLIM only retains the first four characters, this instruction is the same as that used to terminate a macro, but is used in a different context, with different results. Beware of accidentally typing $Endmac twice for a macro; you will lose all of your data, macros, and model definitions.

3. A Symmetry Model

For our second example (Table 1.3), we shall study how subjects self-classify themselves into four social classes: lower, working, middle, or upper class.

Lower	Working	Middle	Upper
72	714	655	41

Table 1.3 Self-Classification of Individuals by Social Class (Haberman, 1978, p.24)

Study of the table shows that many fewer people have chosen the two extreme categories than the central ones. We may ask if this aversion to the extremes is symmetrical for the top and the bottom. Thus, we are interested in determining if the table is symmetric, i.e. if $p_1 = p_4$ and $p_2 = p_3$. This can be translated into a log linear model in the following terms:

$$\begin{aligned} \log(F_k) &= \mu + \alpha \qquad k=1 \text{ or } 4 \\ &= \mu - \alpha \qquad k=2 \text{ or } 3 \end{aligned} \tag{1.9}$$

In this example, and all that follow, the GLIM instructions will be supplied in the corresponding section of Appendix II.

For model (1.9), we require a new vector which may readily be created by the instruction

(28) assign values to a vector - $ASSign name = values separated by commas

This vector has +1 for the first and last values and -1 for the second and third values.

We fit first the equiprobability model followed by the symmetry model. The results from GLIM are as follows:

```
SELF-CLASSIFICATION BY SOCIAL CLASS - HABERMAN (1978, P.24)

 scaled deviance = 1266.8 at cycle  4
           d.f. =    3

 Chi2 probability =  0.      for Chi2 =    1267. with    3. d.f.

         estimate          s.e.      parameter
     1      5.915       0.02594      1
     scale parameter taken as   1.000

   unit  observed     fitted   residual
      1        72     370.50    -15.508
      2       714     370.50     17.846
      3       655     370.50     14.780
      4        41     370.50    -17.118

 scaled deviance = 11.158 at cycle  3
           d.f. =  2

 Chi2 probability =  0.0038 for Chi2 =    11.16 with    2. d.f.

 Chi2 probability =  0.      for Chi2 =    1256. with    1. d.f.

         estimate          s.e.      parameter
     1      5.281       0.04892      1
     2     -1.247       0.04892      CLAS
     scale parameter taken as   1.000

   unit  observed     fitted   residual
      1        72      56.50      2.062
      2       714     684.50      1.128
      3       655     684.50     -1.128
      4        41      56.50     -2.062
```

As may be expected, the equiprobability model fits very badly. The residuals indicate the parabolic form of the relationship. However, the symmetry model, with a Chi-square of 11.16 for 2 d.f., is also to be rejected. The parameter estimate for CLAS is negative, reflecting the fact that fewer people choose the extremes (-1.247 x 1) than the middle (-1.247 x -1). A look at the residuals shows that more people than expected (for this model) classify themselves as lower class as compared to upper class.

This symmetry model may be considered to be a quadratic model centred on the middle of the social class scale:

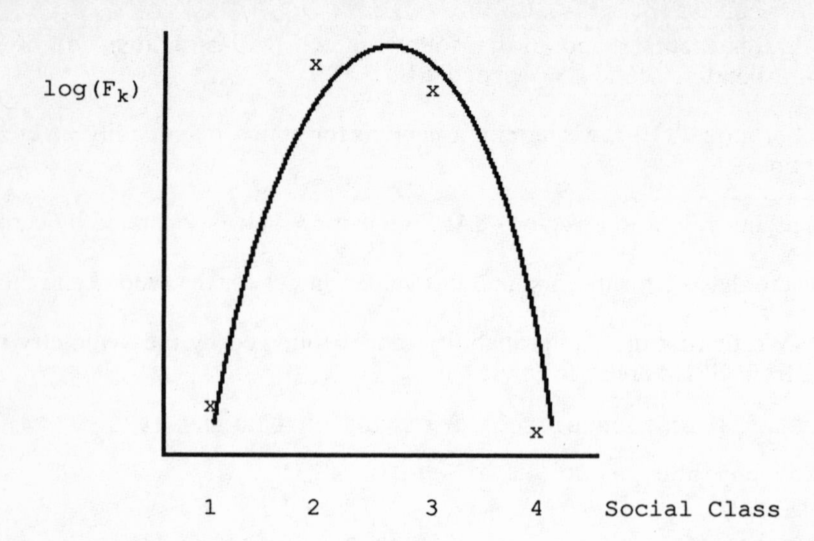

To account for the observed residual differences, it is necessary to shift the parabola, to the left in this case. In other words, we need to add a linear term to the quadratic term already in our model. We shall now have:

$$\log(F_k) = \mu + \beta_1 x_{k1} + \beta_2 x_{k2} \tag{1.10}$$

where $x_{k1} = 2(k-2.5)$ and $x_{k2} = (k-2.5)^2 - 1.25$.

The reader may verify that $\beta_2 x_{k2} = \pm\alpha$ for k=1, 2, 3, 4 so that this parameter remains unchanged between the two models. The choice of 2.5 marks the centre of the scale for 1 to 4. The variables x_{k1} and x_{k2} are called orthogonal polynomials, a subject to which we shall return in Chapter 3.

When we fit this model, we see that the Chi-square test is now satisfactory:

```
SELF-CLASSIFICATION BY SOCIAL CLASS - HABERMAN (1978, P.24)

  scaled deviance = 1.4458 at cycle  3
             d.f. = 1

  Chi2 probability =  0.2292 for Chi2 =   1.446 with   1. d.f.

  Chi2 probability =  0.      for Chi2 =   1265. with   2. d.f.

          estimate          s.e.       parameter
      1      5.271       0.04937       1
      2   -0.06409       0.02066       C1
      3     -1.255       0.04922       C2
      scale parameter taken as  1.000
```

An analysis of residuals is of little use, since only 1 d.f. is left.

4. Periodicity Models

In the first section of this chapter, we have already encountered one simple example of changes in frequency of an event with time. We shall study others in subsequent chapters. However, all such changes are not simple linear trends with time. Just as the days of the week and the seasons show a periodicity, so do many social events. One classical object of sociological study is suicide. Consider, for example, the total number of suicides per month in 1968 for the USA (Table 1.4).

Jan.	Feb.	Mar.	Apr.	May	June	July	Aug.	Sept.	Oct.	Nov.	Dec.
1720	1712	1924	1882	1870	1680	1868	1801	1756	1760	1666	1733

Table 1.4 Suicides in the USA, 1968 (Haberman, 1978, p.51)

A glance at the table shows no systematic pattern: June and November have the least suicides and March the most. We may then first wish to test for the equiprobability of suicide throughout the year. But we must immediately face at least one minor problem: all months do not have the same number of days. The rate per day is the more pertinent statistic to study. However, our log linear model requires absolute frequencies, which rates are not. This factor may, however, be incorporated by including a constant term for days:

$$\log(F_k) = \log(d_k) + \mu \qquad (1.11)$$

where d_k is the number of days in the kth month. The constant term, $\log(d_k)$, is known as an offset, since it does not involve estimation of any unknown parameters. Another similar case would be if we have frequencies of occurrence of an event in various regions or cities, where the latter have different populations. Then the offset incorporates these populations. A new GLIM instruction is required:

(29) introduce a constant term in the linear model - $Offset vector

which will be used in the program included in Appendix II. The results given below show that the equiprobability model must be rejected:

```
SUICIDES, USA, 1968 - HABERMAN (1978, P.51)

  scaled deviance = 37.068 at cycle  3
              d.f. = 11

  Chi2 probability =   0.0001 for Chi2 =    37.07 with    11. d.f.

          estimate          s.e.      parameter
     1       4.067      0.006840         1
     scale parameter taken as     1.000

    unit   observed      fitted    residual
       1       1720       1810.      -2.120
       2       1712       1693.       0.452
       3       1924       1810.       2.675
       4       1882       1752.       3.111
       5       1870       1810.       1.406
       6       1680       1752.      -1.716
       7       1868       1810.       1.359
       8       1801       1810.      -0.216
```

```
  9       1756      1752.      0.100
 10       1760      1810.     -1.180
 11       1666      1752.     -2.050
 12       1733      1810.     -1.814
```

The residuals seem to indicate that there are more suicides in spring and fewer in late autumn and winter, The residual plots (not shown) confirm this.

As a second step, we shall set up a model to allow for differences by season:

$$\log (F_k) = \log (d_k) + \mu + \alpha_j \quad \begin{aligned} j &= 1 \text{ for } k = 1,2,12 \\ &= 2 \text{ for } k = 3,4,5 \\ &= 3 \text{ for } k = 6,7,8 \\ &= 4 \text{ for } k = 9,10,11 \end{aligned} \qquad (1.12)$$

Here, we allow four different probabilities of suicide, one for each season. However, as so described, our model has five parameters instead of the four required. We must add a constraint. This may be done in a number of ways, all of which are mathematically equivalent, but not all of which are as easily interpretable. By default, GLIM sets $\alpha_1 = 0$ so that the other three α_j are comparisons of these three seasons with the first. This is done by defining what is called a factor variable, i.e. a variable with a specific number of nominal levels or categories. These must be numbered from 1 to the maximum. A further instruction tells GLIM which variables, and these may not include the dependent variable, are nominal:

(30) define nominal or factor variables - $FActor series of variables with their numbers of levels

All other variables are taken by GLIM to be measurements and not categorical. However, if necessary, they can be explicitly so defined by

(31) define measured variables - $Variate variable names

For our seasonality model, the results are as follows:

```
SUICIDES, USA, 1968 - HABERMAN (1978, P.51)

 scaled deviance = 12.599 at cycle  3
            d.f. =  8

          estimate          s.e.       parameter
     1       4.039       0.01391       1
     2     0.08341       0.01923       SEAS(2)
     3     0.02408       0.01951       SEAS(3)
     4    0.003286       0.01966       SEAS(4)
     scale parameter taken as   1.000

 Chi2 probability =   0.1256 for Chi2 =   12.60 with   8. d.f.

 Chi2 probability =   0.0000 for Chi2 =   24.47 with   3. d.f.
```

By the introduction of three new parameters, we obtain a very significant reduction in the Chi-square. And the remaining Chi-square is not significant. We see that significantly more suicides occurred in the spring and less in autumn and winter, while summer is in between: for winter (category 1), the estimate is 4.039 and for autumn (4.039 + 0.003 =) 4.042, while for spring, it is (4.039 + 0.083 =) 4.122. The residuals and plots (not shown) no longer indicate any clear trend. Note that, while residual tables and plots will not always be included in the text, for lack of space, they should always be produced and inspected for regularities, if the degrees of freedom are not too small.

An alternative, but equivalent, way of placing a constraint on our model is to have

$$\Sigma\alpha_j = 0$$

This is known as the conventional constraint and provides us with comparisons around the mean instead of with respect to one privileged category, the first, as was the case in what just preceded. However, this is more complex with GLIM, since it cannot be done automatically. Instead, what is known as a design matrix must be defined. This matrix is, in fact, a series of vectors, one for each parameter to be estimated. In our case, we have three parameters, the fourth being given by the sum to zero. To simplify matters, a general macro, called TRAN, is provided in Appendix III, which generates such vectors for any factor variable with up to 9 levels. This macro is loaded in the same way as CHIT and RESP, as can be seen from the program for this section in Appendix II. The results are:

```
SUICIDES, USA, 1968 - HABERMAN (1978, P.51)

 scaled deviance = 12.599 at cycle  3
            d.f. =  8

          estimate         s.e.      parameter
      1      4.066     0.006845      1
      2   -0.02769      0.01199      S1
      3    0.05572      0.01162      S2
      4  -0.003618      0.01185      S3
      scale parameter taken as  1.000
```

We note immediately that the deviance is identical to the case when we used $FActor. This should not be surprising since we are fitting the same model, but simply with different constraints. We next note that the differences between seasons are the same in the two cases. For example, in the first case, the contrast between spring and winter is given directly as 0.08341, while here it is 0.05572 - (-0.02769) = 0.08341. Thus, our interpretation does not change. In more complicated examples in the subsequent chapters, this second set of constraints will prove invaluable as an aid to interpretation.

It is evident that we have here a cyclical phenomenon, but the choice of seasons as the period for the cycles may seem arbitrary for suicides. A more abstract and neutral model may be constructed using trigonometrical functions:

$$\log(F_k)=\log(d_k)+\beta_0 + \beta_1\sin[(2k-1)\pi/12] + \beta_2\cos[(2k-1)\pi/12] \quad (1.13)$$

GLIM has a function for the sine, %SIN(), but none for the cosine, so that we may use the relationship $\sin^2 a + \cos^2 a = 1$. We note that this model has one less parameter than the seasonality one: two parameters in addition to the mean. The results are

```
SUICIDES, USA, 1968 - HABERMAN (1978, P.51)

 scaled deviance = 14.695 at cycle  3
            d.f. =  9
```

```
Chi2 probability =  0.0990 for Chi2 =    14.70 with    9. d.f.

Chi2 probability =  0.0000 for Chi2 =    22.37 with    2. d.f.

          estimate            s.e.       parameter
1             4.114        0.01638       1
2           0.03376       0.009616       SIN
3          -0.07312        0.02331       COS
scale parameter taken as   1.000
```

As in the first time trend model, a plot of observed and fitted values is also useful here. The first plot shows the observed and fitted frequency of suicides over the twelve months. The second plot, obtained by taking logarithms of the observed and fitted values, shows the sine-cosine curve itself.

```
SUICIDES, USA, 1968 - HABERMAN (1978, P.51)

Observed and fitted values

  1952.0 |
  1936.0 |
  1920.0 |          2
  1904.0 |
  1888.0 |              F
  1872.0 |                  F       F
  1856.0 |              %
  1840.0 |                  %
  1824.0 |
  1808.0 |                              F       %
  1792.0 |
  1776.0 |  %
  1760.0 |                          %   %   F   F       %
  1744.0 |                                  %
  1728.0 |  F   %           %                           F
  1712.0 |      F
  1696.0 |                                          %
  1680.0 |                  F
  1664.0 |                                          F
  1648.0 |
 --------:---------:---------:---------:---------:---------:--------
       0.00      2.50      5.00      7.50     10.00     12.50     15.

Harmonic model

 7.56800 |
 7.56000 |          2
 7.55200 |
 7.54400 |              F
 7.53600 |                  F       F
 7.52800 |              T
 7.52000 |                  T
 7.51200 |
 7.50400 |
 7.49600 |                              F       T
 7.48800 |  T
 7.48000 |
 7.47200 |                          T   T   F   F       T
 7.46400 |                                  T
 7.45600 |      T           T                           F
 7.44800 |  F   F
 7.44000 |                                          T
 7.43200 |
 7.42400 |                  F
 7.41600 |                                          F
 --------:---------:---------:---------:---------:---------:--------
       0.00      2.50      5.00      7.50     10.00     12.50     15.
```

Although the deviance is somewhat higher than in the seasonality model, this is compensated by the gain of 1 d.f. The largest deviations between observations and model on the graphs occur in early summer, especially in June. This is confirmed by the residuals and plots (not shown). However, both models, the seasonal and the harmonic, do fit the data well. The test will be to apply them to similar data for other years to see which one maintains a reasonable fit in varying circumstances.

5. Local Effects

We shall now consider one final simple example with a one-way frequency table. At times, a model fits only a part of the observations and the rest must be ignored in constructing the model. Durkheim (1897, p.101) studied the suicide rate per day (Table 1.5).

Monday	Tuesday	Wednesday	Thursday	Friday	Saturday	Sunday
1001	1035	982	1033	905	737	894

Table 1.5 Durkheim's Suicides (Haberman, 1978, p.87)

In this table, more suicides seem to occur at the beginning of the week than at the end. We first test for equiprobability of suicide for all days of the week:

```
SUICIDES (DURKHEIM) - HABERMAN (1978, P.87)

 scaled deviance = 74.918 at cycle  3
            d.f. =  6

 Chi2 probability =  0.      for Chi2 =    74.92 with    6. d.f.

         estimate          s.e.       parameter
     1       6.847      0.01232        1
     scale parameter taken as   1.000

   unit  observed      fitted    residual
      1      1001       941.0       1.956
      2      1035       941.0       3.064
      3       982       941.0       1.337
      4      1033       941.0       2.999
      5       905       941.0      -1.174
      6       737       941.0      -6.650
      7       894       941.0      -1.532
```

```
SUICIDES (DURKHEIM) - HABERMAN (1978, P.87)

Poisson Residuals

Score Test Coefficient of Sensitivity

    0.000 I
   -0.500 I  S                S                S               S
   -1.000 I
   -1.500 I                           S
   -2.000 I           S
   -2.500 I
   -3.000 I
   -3.500 I
   -4.000 I
   -4.500 I
   -5.000 I
   -5.500 I
   -6.000 I
   -6.500 I
   -7.000 I
   -7.500 I
   -8.000 I
   -8.500 I                                            S
   -9.000 I
   -9.500 I
----------:---------:---------:---------:---------:---------:--------
        0.60      1.80      3.00      4.20      5.40      6.60     7.

Residual Plot

    3.600 I                                                     A
    3.000 I                                          2          S
    2.400 I                                   A
    1.800 I                                   S
    1.200 I                             2                       Y
    0.600 I                                   Y      Y
    0.000 I                             Y
   -0.600 I                Y      Y
   -1.200 I     Y                 2
   -1.800 I                2
   -2.400 I
   -3.000 I
   -3.600 I
   -4.200 I
   -4.800 I
   -5.400 I
   -6.000 I
   -6.600 I     S
   -7.200 I     A
   -7.800 I
----------:---------:---------:---------:---------:---------:--------
       -1.800    -1.200    -0.600    0.000     0.600     1.200     1.8

Points Y represent 45 line
```

The model is clearly rejected. We see from the residuals that it underestimates the rate during the first four weekdays and overestimates that for Friday and the weekend. Saturday especially stands out, with a much lower rate than any other day.

In a second model, we construct a binary variable, with four ones and three minus ones, to distinguish between these two periods of the week:

```
SUICIDES (DURKHEIM) - HABERMAN (1978, P.87)

 scaled deviance = 23.351 at cycle  3
            d.f. =  5

 Chi2 probability =  0.0003 for Chi2 =   23.35 with   5. d.f.

 Chi2 probability =  0.     for Chi2 =   51.57 with   1. d.f.
```

```
        estimate          s.e.        parameter
    1      6.830       0.01266        1
    2      0.09035     0.01266        DAYS
    scale parameter taken as   1.000

  unit  observed     fitted   residual
     1      1001     1012.7     -0.369
     2      1035     1012.7      0.699
     3       982     1012.7     -0.966
     4      1033     1012.7      0.636
     5       905      845.3      2.052
     6       737      845.3     -3.726
     7       894      845.3      1.674
```

```
SUICIDES (DURKHEIM) - HABERMAN (1978, P.87)

Poisson Residuals

Score Test Coefficient of Sensitivity

     0.000 |   S        S             S
    -0.600 |                   S
    -1.200 |
    -1.800 |
    -2.400 |                                               S
    -3.000 |                                S
    -3.600 |
    -4.200 |
    -4.800 |
    -5.400 |
    -6.000 |
    -6.600 |
    -7.200 |
    -7.800 |
    -8.400 |
    -9.000 |
    -9.600 |
   -10.200 |                                       S
   -10.800 |
   -11.400 |
 ---------:---------:---------:---------:---------:---------:--------
        0.60      1.80      3.00      4.20      5.40      6.60      7.

Residual Plot

     2.800 |
     2.400 |
     2.000 |                                                    A
     1.600 |                                        A           S
     1.200 |                                        S           Y
     0.800 |
     0.400 |                              2     2       Y
     0.000 |                                    Y
    -0.400 |                              Y
    -0.800 |                         3
    -1.200 |                 2
    -1.600 |                 A
    -2.000 |      Y
    -2.400 |
    -2.800 |
    -3.200 |
    -3.600 |
    -4.000 |      S
    -4.400 |
    -4.800 |      A
 ---------:---------:---------:---------:---------:---------:--------
      -1.800    -1.200    -0.600     0.000     0.600     1.200     1.8

Points Y represent 45 line
```

Now, the residuals and plots indicate that reasonable estimates appear to be given for the four weekdays, but Friday and the weekend, especially Saturday, still pose a problem. The model is not yet acceptable.

Let us then ignore completely these three days, which seem to vary among themselves, and fit an equiprobability model to the four weekdays. To do this, we can introduce a weighting factor, with unit weight for the days of interest and zero weight for the others. The necessary GLIM instruction is:

(32) define weights for the observations - $Weight weight vector

Such an instruction may also be used for grouped frequency data, where combined observations are given weights (greater than unity) equal to their observed numbers.

We construct a vector with four ones and three zeroes as the weight and refit the equiprobability model:

```
SUICIDES (DURKHEIM)  - HABERMAN (1978, P.87)

 -- model changed
 scaled deviance = 1.9676 at cycle  3
            d.f. = 3        from 4 observations

 Chi2 probability =  0.5831 for Chi2 =    1.968 with    3. d.f.

          estimate        s.e.     parameter
     1       6.920     0.01571      1
     scale parameter taken as   1.000

   unit   observed      fitted   residual
     1       1001      1012.8     -0.369
     2       1035      1012.8      0.699
     3        982      1012.8     -0.966
     4       1033      1012.8      0.636
     5        905      1012.8      0.000
     6        737      1012.8      0.000
     7        894      1012.8      0.000
```

Our model now fits very well. Suicide is equally probable on the first four weekdays but varies among Friday, Saturday, and Sunday. We see that the residuals are zero for these three days, so that we are, in fact, fitting the model exactly to these three observations (although GLIM prints out the general mean as the fitted value).

Through this series of simple frequency tables, we have now encountered many of the basic principles of analysis with GLIM, as well as all of the necessary instructions to be used in the following chapters. The only major new aspects of GLIM still to be introduced are certain GLIM macros for special applications. These macros, as those more general ones already introduced, CHIT, RESP, and TRAN, will be found in Appendix III.

We are ready to proceed to more complex models involving frequencies classified by several variables.

CHAPTER 2

TIME AND CAUSALITY

1. Retrospective Studies I

An eternal problem in the social sciences is that of determining direction of causality. This is not unique to these disciplines, since it also appears with the same force, for example, in the medical sciences or in astronomy. Most of the natural sciences are able to resolve the problem through the application of experimental methods. Such is not possible for the human sciences. Thus, no unequivocal means is available to determine causality where experimentation must be excluded. Various indirect methods must be applied. The more mutually-confirming approaches used, the more confident may we become that we are perhaps succeeding in isolating a cause.

One of the most useful approaches to the problem of studying causality in this context is the use of a time factor. Events which occur later in time cannot affect earlier events, or at least we may so suppose in many case. Two approaches to collecting chronological information may be distinguished: (1) we may choose a sample of individuals according to the criteria of certain explanatory variables and then follow them up in time to see what response variable, the variable to be explained, is obtained or (2) we may choose a sample according to the response variable and then investigate what values of the explanatory variables had previously (in time) existed. The first case is a prospective study. It includes panel studies and cohort studies. We shall consider it in later sections of this chapter. The second is a retrospective study. In the medical sciences, it is often called a case control study. A common example in the social sciences is the study of social mobility to which we now turn. We may note that the first approach resembles experimentation in the natural sciences, with, however, absence of random allocation of the explanatory variables. And, in fact, the methods of statistical analysis are often identical, although the strength of conclusions cannot be. In contrast, the second approach is specific to the human sciences and often requires special analytic procedures.

In a social mobility study, as in many related social studies, we obtain a sample of people with their characteristics, and then retrospectively obtain information about their parents. For social mobility, the information is specifically about occupation, but the same principle applies for education, political beliefs, and so on.

As the name implies, a retrospective study does things backwards. We have a certain number of children (almost invariably sons) of each occupational category, and we look back to see from which parental (father's) occupational category they came. We have a sampling structure which implies that we can calculate the probability of the father having any given occupation given the son's occupation. This is the exact opposite of what we want. In addition, our sample, if correctly chosen, will be representative of the sons' occupations but not of the fathers'. The occupational structure may have changed between the two generations, but we are interested in mobility, not in these structural changes.

Our observations take the form of a two-way table, cross- classifying the two occupational variables. Our two problems, the retrospective nature of the study and the changes in occupational structure, may be resolved by the same procedure: we study changes within the table given (conditional on the fact) that the marginal totals are fixed; we then apply the log linear model. It is possible to demonstrate that this is the only procedure which can resolve these two problems.

Son 1	2	3	4	5	Father	Categories
50	45	8	18	8	1	Professional, High Administrative
28	174	84	154	55	2	Managerial, Executive, High Supervisory
11	78	110	223	96	3	Low Inspectional, Supervisory
14	150	185	714	447	4	Routine Nonmanual, Skilled Manual
3	42	72	320	411	5	Semi- and Unskilled Manual

Table 2.1 British Inter-Generational Social Mobility (Glass, 1954, as modified by Bishop et al, 1975, p.100)

We shall now briefly consider a classical social mobility table, derived from Glass (1954): Table 2.1. More detailed analyses of such mobility tables will be delayed until Chapters 5 and 6, where a number of specific models for such studies will be introduced. Here we shall only consider whether or not the son's occupation depends on the father's. If it does not, the two variables are said to be independent. For this example, we shall again fit the model in two ways, with factor variables ($FActor) and by constructing our own design matrix.

With GLIM, a cross-tabulated table, to which log linear models are to be fitted, is stored in a single vector containing the observed frequencies. A series of other vectors must be defined to index the row, columns, and so on, of the multi-dimensional matrix. Each of these latter vectors represents a variable to be fitted to the data. Thus, analysis of log linear models with GLIM involves one more vector of values than the number of variables to be included in the model. Here, our two-way table has two variables, father's and son's occupation, but requires three vectors.

We first fit the model where only the two sets of marginal frequencies are fixed:

$$\log(F_{ik}) = \mu + \theta_i + \varnothing_k \tag{2.1}$$

This is the model for independence between the two occupational situations. Each of the (mean) parameter vectors is analogous to those already encountered in Chapter 1. The results show that this model cannot be accepted:

```
BRITISH SOCIAL MOBILITY - GLASS (1954)

 scaled deviance = 792.19 at cycle  5
           d.f. =  16
```

	estimate	s.e.	parameter
1	1.363	0.1300	1
2	1.529	0.1071	SON(2)
3	1.466	0.1077	SON(3)
4	2.601	0.1006	SON(4)
5	2.261	0.1020	SON(5)
6	1.345	0.09884	FATH(2)
7	1.390	0.09839	FATH(3)
8	2.460	0.09172	FATH(4)
9	1.883	0.09449	FATH(5)

scale parameter taken as 1.000

BRITISH SOCIAL MOBILITY - GLASS (1954)

scaled deviance = 792.19 at cycle 5
d.f. = 16

	estimate	s.e.	parameter
1	4.350	0.02949	1
2	-1.571	0.07919	SON1
3	-0.04248	0.04290	SON2
4	-0.1058	0.04382	SON3
5	1.030	0.03215	SON4
6	-1.416	0.07205	FAT1
7	-0.07086	0.04188	FAT2
8	-0.02544	0.04123	FAT3
9	1.044	0.03064	FAT4

scale parameter taken as 1.000

unit	observed	fitted	residual
1	50	3.907	23.320
2	45	18.023	6.354
3	8	16.917	-2.168
4	18	52.669	-4.777
5	8	37.484	-4.816
6	28	14.991	3.360
7	174	69.159	12.607
8	84	64.916	2.369
9	154	202.101	-3.384
10	55	143.833	-7.407
11	11	15.688	-1.184
12	78	72.372	0.662
13	110	67.932	5.104
14	223	211.492	0.791
15	96	150.516	-4.444
16	14	45.731	-4.692
17	150	210.969	-4.198
18	185	198.026	-0.926
19	714	616.511	3.926
20	447	438.763	0.393
21	3	25.682	-4.476
22	42	118.478	-7.026
23	72	111.209	-3.718
24	320	346.226	-1.409
25	411	246.405	10.486

```
BRITISH SOCIAL MOBILITY - GLASS (1954)

Poisson Residuals

Score Test Coefficient of Sensitivity

      0.00 I     S     S   S     S S   S       S   S
    -10.00 I   S                     S               S   S S
    -20.00 I         S       S           S
    -30.00 I       S                       S
    -40.00 I S                               S         S
    -50.00 I
    -60.00 I                   S
    -70.00 I
    -80.00 I             S
    -90.00 I                                     S
   -100.00 I
   -110.00 I
   -120.00 I
   -130.00 I
   -140.00 I
   -150.00 I
   -160.00 I
   -170.00 I
   -180.00 I                                                 S
   -190.00 I
 ----------:---------:---------:---------:---------:---------:---------
         0.00      5.00     10.00     15.00     20.00     25.00     30.

Residual Plot

     26.00 I
     24.00 I                                                2
     22.00 I
     20.00 I
     18.00 I
     16.00 I
     14.00 I                                        A  A
     12.00 I                                           S
     10.00 I                                        S
      8.00 I
      6.00 I                                   A2 2
      4.00 I                                 2 S
      2.00 I                               A2     Y Y  Y    Y
      0.00 I                   YY YYYYYYY2332YY YY
     -2.00 I       Y    Y  Y Y        222A
     -4.00 I               S S SS SS22
     -6.00 I               A A AA AA
     -8.00 I       S    2
    -10.00 I       A
    -12.00 I
 ----------:---------:---------:---------:---------:---------:---------
        -3.00     -2.00     -1.00      0.00      1.00      2.00      3.

Points Y represent 45 line
```

The size of the Chi-square is such that it is not even necessary to check the level of significance. A glance at the residuals and at the coefficient of sensitivity shows that the diagonal residuals are large and positive; the model has underestimated these values. This model of independence does not predict the observed fact that many sons remain in the same occupational category as their fathers.

An alternative is to fit what is known as the saturated model, a model with as many parameters as there are entries in the table:

$$\log (F_{ik}) = \mu + \theta_i + \phi_k + \gamma_{ik} \qquad (2.2)$$

This model must necessarily fit the data exactly. Here γ_{ik} is a matrix of parameters describing the mobility between generations under the conditions set forth above: a retrospective study with changing occupational structure. As with the mean parameter vectors, constraints must be applied to this parameter matrix in order to be able to estimate the model. As with factor variables in Chapter 1, GLIM very simply sets the first row and the first column to zero so that all remaining values are comparisons with the first category of each variable. Here the complications of interpretation begin. But the model may very easily be fitted by placing a dot between the two variable names which have been declared in $FActor:

```
$Fit FATH + SON + FATH.SON
```

The three terms correspond exactly to the last three in the log linear model (2.2) above. GLIM implicitly fits the mean. Another equivalent formulation is

```
$Fit FATH * SON
```

The use of the asterix implies that all lower order terms are automatically included. Note that this is not the same as multiplying FATH*SON in $CAlculate and then fitting the result. The latter adds only one (linear interaction) term to the model, while the former adds a large number of main effect and interaction terms.

In contrast to this, construction of a design matrix is considerably more complex. Each of the two variables must be translated into a series of variables (four in this case) using TRAN. Then, all possible products of these two sets of four variables, i.e. 16 new variables, must be calculated. This is possible by means of macros, but unfortunately, a different macro is required for each size of variable encountered (i.e. 2x2, 2x3, 2x4, ...). One example of such a macro is given in Appendix III. Note also the insertion of the resulting term in the fit by means of a second macro with #.

The results from GLIM by the two approaches are:

```
BRITISH SOCIAL MOBILITY - GLASS (1954)

 scaled deviance = 0.00000000 at cycle  3
             d.f. = 0

         estimate       s.e.      parameter
   1       3.912       0.1414     1
   2     -0.1054       0.2055     SON(2)
   3      -1.833       0.3808     SON(3)
   4      -1.022       0.2749     SON(4)
   5      -1.833       0.3808     SON(5)
   6     -0.5798       0.2360     FATH(2)
   7      -1.514       0.3330     FATH(3)
   8      -1.273       0.3024     FATH(4)
   9      -2.813       0.5944     FATH(5)
  10       1.932       0.2893     SON(2).FATH(2)
  11       2.064       0.3820     SON(2).FATH(3)
  12       2.477       0.3469     SON(2).FATH(4)
  13       2.744       0.6319     SON(2).FATH(5)
  14       2.931       0.4389     SON(3).FATH(2)
  15       4.135       0.4950     SON(3).FATH(3)
  16       4.414       0.4710     SON(3).FATH(4)
  17       5.011       0.7016     SON(3).FATH(5)
  18       2.726       0.3432     SON(4).FATH(2)
  19       4.031       0.4135     SON(4).FATH(3)
  20       4.953       0.3852     SON(4).FATH(4)
  21       5.691       0.6419     SON(4).FATH(5)
  22       2.508       0.4460     SON(5).FATH(2)
```

```
23        3.999        0.4963     SON(5).FATH(3)
24        5.296        0.4676     SON(5).FATH(4)
25        6.753        0.6934     SON(5).FATH(5)
  scale parameter taken as    1.000

BRITISH SOCIAL MOBILITY - GLASS (1954)

 scaled deviance = 0.00000000 at cycle  3
         d.f. = 0

         estimate          s.e.      parameter
 1        4.184       0.04034      1
 2       -1.508        0.1219      SON1
 3       0.2300       0.05778      SON2
 4     -0.04262       0.07480      SON3
 5       0.9506       0.05773      SON4
 6       -1.231       0.09987      FAT1
 7       0.2091       0.05917      FAT2
 8       0.1012       0.06850      FAT3
 9       0.9245       0.06098      FAT4
10        2.467        0.1744      RR11
11       0.4471        0.1720      RR12
12      -0.3792        0.2251      RR13
13      -0.9614        0.2066      RR14
14       0.6231        0.1403      RR21
15       0.5358       0.08532      RR22
16      -0.1586        0.1050      RR23
17      -0.3280       0.08848      RR24
18      -0.8315        0.2428      RR31
19      0.08016        0.1084      RR32
20       0.4578        0.1092      RR33
21       0.1543       0.09814      RR34
22       -1.014        0.1780      RR41
23      -0.3069       0.08685      RR42
24       0.1713       0.08951      RR43
25       0.5116       0.07700      RR44
  scale parameter taken as    1.000
```

For this second approach, using the design matrix, the parameter values for the last line and column may be obtained through the constraint that each line or column sums to zero: the value is obtained by adding up the row or column and changing the sign. Here the matrix is

```
  2.467     0.4471   -0.3792   -0.9614   -1.5735
  0.6231    0.5358   -0.1586   -0.3280   -0.6723
 -0.8315    0.0802    0.4578    0.1543    0.1392
 -1.014    -0.3069    0.1713    0.5116    0.6380
 -1.2446   -0.7562   -0.0913    0.6235    1.4686
```

The deviance for this saturated model is zero, corresponding to the fact that it fits perfectly. In this case, study of the residuals is of no use, since they are all zero.

We see from our parameter matrix once again how the diagonal categories are over-represented. Members of the two extreme categories, professional and high administrative and semi- and unskilled have especially little mobility. We now have values which eliminate the bias from the retrospective method and from changes in occupational structure.

In Chapters 5 and 6, we shall study other intermediate models between the independence and saturated models for such square tables as this mobility table.

2. Retrospective Studies II

Our second example of a retrospective study is a typical case- control study (Table 2.2). The response variable of interest is use of a university contraceptive clinic. A sample of clinic users comprised the case group and a corresponding sample of non-users formed the control. The data were checked to verify that the two groups were similar for various background variables. Most often in case-control studies, the individuals are actually matched on these variables.

		Use Clinic	
Virgin	**Attitude to Sex**	Yes	No
Yes	Always Wrong	23	23
Yes	Not Always Wrong	29	67
No	Always Wrong	127	18
No	Not Always Wrong	112	15

Table 2.2 Clinic Use, Attitude to Extra-Marital Sex, and Virginity (Fienberg, 177,p.92)

The explanatory variables of interest are virginity and attitude to extra-marital sexual relations. Each of the three variables involved is binary or dichotomous. When the response variable is binary, analysis with GLIM may be simplified by using a binomial distribution instead of Poisson. For this example, we shall present the two approaches to demonstrate that the results are the same and to illustrate the relationships between them. We begin with the analysis using the Poisson distribution, that is with the log linear model:

$$\log (F_{ijk}) = \mu + \theta_i + \phi_j + \omega_k + \alpha_{ik} + \beta_{jk} + \psi_{ij} + \gamma_{ijk} \qquad (2.3)$$

We consider here only the conventional constraints of summation to zero and shall not use the GLIM $FActor facility. For dichotomous variables, the design matrix can be constructed very simply without need for the macro TRAN, as seen in the program in Appendix II.

We now have a much more complex model than previously since, with more than two variables present, they can interact in groups of two or more. The indices indicate which variables are interacting.

For the log linear model, we must consider that the marginal totals for the three variables are fixed, as are the totals for the relationship between the two explanatory variables, attitude and virginity. Thus, our base model is

$$\log (F_{ijk}) = \mu + \theta_i + \phi_j + \omega_k + \psi_{ij} \qquad (2.4)$$

After fitting, we see that this base model must be rejected:

```
CLINIC USE (FIENBERG, 1976 P.92)

 scaled deviance = 121.34 at cycle  4
           d.f. =   3

 Chi2 probability =  0.      for Chi2 =    121.3 with   3. d.f.

         estimate          s.e.      parameter
     1      3.770       0.05837      1
     2    -0.1508       0.05415      ATTI
     3    -0.3570       0.05415      VIRG
     4     0.4306       0.05376      USE
     5    -0.2171       0.05415      AV
     scale parameter taken as  1.000
```

Clinic use depends either upon attitude or upon virginity or both. We now introduce the relationship between clinic use and virginity and immediately find that we have an acceptable model:

```
CLINIC USE (FIENBERG, 1976 P.92)

 scaled deviance = 5.1905 (change =  -116.1) at cycle  3
           d.f. = 2       (change =      -1  )

 Chi2 probability =  0.0746 for Chi2 =    5.190 with   2. d.f.

 Chi2 probability =  0.      for Chi2 =    116.1 with   1. d.f.

         estimate          s.e.      parameter
     1      3.629       0.06560      1
     2    -0.1508       0.05415      ATTI
     3    -0.1625       0.06560      VIRG
     4     0.3578       0.06365      USE
     5    -0.2171       0.05415      AV
     6    -0.6321       0.06365      UV
     scale parameter taken as   1.000
```

Virgins have a lower probability of using the clinic, as indicated by the negative value of the parameter for the use/virginity interaction. If, instead of this relationship, we substitute that between clinic use and attitude, we see that the model is not acceptable:

```
CLINIC USE (FIENBERG, 1976 P.92)

 scaled deviance = 109.60 (change =  +104.4) at cycle  4
           d.f. =   2     (change =       0  )

 Chi2 probability =  0.0000 for Chi2 =    109.6 with   2. d.f.

 Chi2 probability =  0.0006 for Chi2 =    11.73 with   1. d.f.

         estimate          s.e.      parameter
     1      3.744       0.06040      1
     2    -0.2312       0.06040      ATTI
     3    -0.3570       0.05415      VIRG
     4     0.4598       0.05608      USE
```

```
5        -0.2171         0.05415         AV
6         0.1888         0.05608         UA
scale parameter taken as    1.000
```

Finally, if we include both relationships at the same time, we again have an acceptable model, but not significantly better than that with only the relationship between clinic use and virginity.

```
scaled deviance = 2.9165 (change =   -106.7) at cycle   3
           d.f. = 1        (change =      -1  )

Chi2 probability =   0.0877 for Chi2 =    2.916 with    1. d.f.

Chi2 probability =   0.0000 for Chi2 =    118.4 with    2. d.f.


         estimate          s.e.        parameter
1          3.629        0.06548        1
2        -0.1765        0.05727        ATTI
3        -0.1593        0.06541        VIRG
4         0.3746        0.06513        USE
5        -0.1680        0.06286        AV
6        0.09879        0.06553        UA
7        -0.6162        0.06428        UV
scale parameter taken as    1.000
```

Thus, we can exclude the relationship between clinic use and attitudes to sex outside marriage.

We now repeat the analysis using the binomial distribution. Our model is now

$$\log (F_{ij1}/F_{ij2}) = \omega + \alpha_i + \beta_j + \gamma_{ij} \qquad (2.5)$$

where the four parameter sets correspond to those with the same Greek letters in model (2.3) above. With the model presented in this way, we may interpret the relationship between the three variables at the same time (gamma), as the statistical interaction between the two explanatory variables with respect to the response variable (in the same way as for normal theory ANOVA or regression).

In the case of the binomial distribution, $ERror requires specification of an additional vector, that of the binomial denominator. The data must be presented somewhat differently to those for the log linear model above, since we now have two vectors containing the observed frequencies, here users and the total for each combination of categories of the explanatory variables. This model with a binary response is often known as the logistic model.

If we follow the same steps as above, GLIM gives the following results:

```
CLINIC USE (FIENBERG, 1976 P.92)

 scaled deviance = 121.34 at cycle   3
            d.f. =    3

 Chi2 probability =  0.      for Chi2 =    121.3 with    3. d.f.
```

```
          estimate          s.e.      parameter
     1      0.8611        0.1074      1
     scale parameter taken as   1.000

 scaled deviance = 5.1905 (change =   -116.1) at cycle   3
           d.f. = 2       (change =      -1  )

 Chi2 probability =  0.0746 for Chi2 =    5.190 with   2. d.f.

 Chi2 probability =  0.       for Chi2 =    116.1 with   1. d.f.

          estimate          s.e.      parameter
     1      0.7157        0.1273      1
     2     -1.264         0.1273      VIRG
     scale parameter taken as   1.000

CLINIC USE (FIENBERG, 1977, P.92)

 scaled deviance = 109.60 (change =   +104.4) at cycle   3
           d.f. =   2     (change =       0  )

 Chi2 probability =  0.0000 for Chi2 =    109.6 with   2. d.f.

 Chi2 probability =  0.0006 for Chi2 =    11.73 with   1. d.f.

          estimate          s.e.      parameter
     1      0.9195        0.1119      1
     2      0.3775        0.1119      ATTI
     scale parameter taken as   1.000

 scaled deviance = 2.9165 (change =   -106.7) at cycle   3
           d.f. = 1       (change =      -1  )

 Chi2 probability =  0.0877 for Chi2 =    2.916 with   1. d.f.

 Chi2 probability =  0.0000 for Chi2 =    118.4 with   2. d.f.

          estimate          s.e.      parameter
     1      0.7492        0.1302      1
     2      0.1976        0.1310      ATTI
     3     -1.232         0.1285      VIRG
     scale parameter taken as   1.000
```

We see that all of the deviances (and Chi-squares) are the same as for the log-linear model but that the parameter values of interest are twice as large for the logistic model (2.5) as for the log linear model (2.3). Model fitting is simpler and more efficient because the base model is the minimal model and the vectors are half as long in this case.

We have analysed this example as if the relationship between response and explanatory variable were clear. A moment's reflection, however, shows that no specific time sequence necessarily holds among the variables. Any one of the three might be the response, explained by the other two. Indeed, our first analysis, with the log linear model, permits any one of the three interpretations, although the analysis would proceed slightly differently due to choice of the base model and the subsequent inclusion of the relationship between a different pair of variables in it. Or all three variables may be considered on the same level and their inter-relationships studied. In this case, only the three main effects are included in the base model. Again, the log linear model permits this. Response and explanatory variables are distinguished only by the interpretation of the analyst and not by the statistics of the analysis. This is directly related to the fact that the log linear (and logistic) model may be equally well applied to prospective and retrospective studies.

Our final model must include the relationships between attitude and virginity (the reader is invited to verify that this is necessary) and between virginity and clinic use. At least six distinct lines of causality may be imagined:

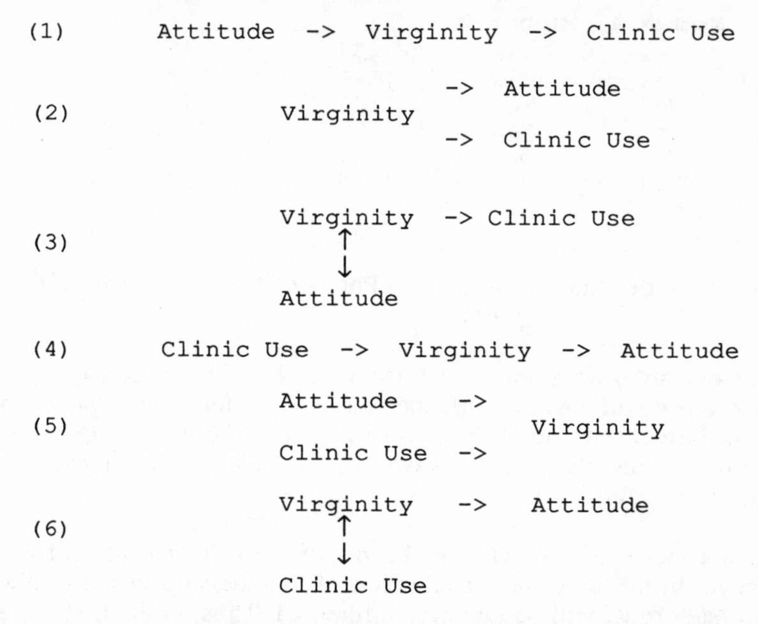

Here, a single-headed arrow indicates causality and a double-headed arrow a non-causal relationship of association. In this series of causal models, we have not even considered the possibility that some other factor, not included in the study, personality, social class, biological makeup, etc., is, in fact, the underlying causal factor for two or all three of these variables.

In no case, whether choice among the six or more alternative causal models or consideration of an external factor, can the analysis with the log linear model aid in making the decision. The choice must depend on other social information, perhaps combined with knowledge of the way the data were collected. *Post hoc* statistical analysis cannot resolve problems of causality.

3. **Panel Studies**

A panel study involves posing the same set of questions to the same individuals at several points in time. Here, we necessarily avoid the time-sequential ambiguity of the previous example, at least for relations between time points. The ambiguity may still remain for relationships among responses at the same time point. In this section, we shall consider a classical example of a two-wave panel with two variables at each time point. This is a study of schoolboys, their membership in the "leading crowd", and their attitude to it (Table 2.3).

As in the previous example, all variables are dichotomous, so that we may use the logistic model with the binomial distribution. This involves entering the table three times, after collapsing it over various variables.

			Attitude 2	
Attitude 1	Member 1	Member 2	Favourable	Unfavourable
Favourable	Yes	Yes	458	140
Unfavourable	Yes	Yes	171	182
Favourable	No	Yes	184	75
Unfavourable	No	Yes	85	97
Favourable	Yes	No	110	49
Unfavourable	Yes	No	56	87
Favourable	No	No	531	281
Unfavourable	No	No	338	554

Table 2.3 Members of the Leading Crowd at Two Points in Time (Coleman, 1964, p.171)

In the previous example, we studied all three variables simultaneously in the same model, since any one could possibly influence the others. Here, the responses of the second time cannot influence those of the first so that we must analyze them in a model which does not include the second wave (i.e. a table collapsed over the variables of the second time point).

In general, in a panel study, it will not be possible to distinguish order of causality within a wave. In our case, self- evaluation of membership in the leading crowd and attitude to that crowd will be mutually influential. Thus, in the first wave, we take attitude towards the leading crowd and membership as mutually interacting. The relationship is very significant. Members are more favourable to the existence of a leading crowd than are non-members:

```
MEMBERS OF THE LEADING CROWD (COLEMAN, 1964, P.171)

 Response variable: ATT1

 scaled deviance = 35.163 at cycle  3
            d.f. =  1

 Chi2 probability =  0.      for Chi2 =   35.16 with   1. d.f.

         estimate        s.e.      parameter
    1     0.1521      0.03441      1
    scale parameter taken as    1.000

  unit  observed     out of    fitted    residual
    1        757       1253     674.1       4.699
    2       1071       2145    1153.9      -3.592

 scaled deviance = 0.0000000 (change =   -35.16) at cycle  3
            d.f. = 0         (change =    -1    )
```

```
        estimate          s.e.       parameter
    1      0.2100       0.03606      1
    2      0.2128       0.03606      MEM
    scale parameter taken as   1.000
```

Note that, with one degree of freedom, study of residuals is not informative.

We shall now rather arbitrarily take a first variable of the second wave, membership, as it depends on previous membership and attitude at the first time point. We fit the model with neither explanatory variable, then that with membership at the first wave added. The Chi-square is very greatly reduced, as might be expected with two so closely related variables, but a significant lack of fit still remains. We, then, add the attitude variable and obtain a very good fit of the model:

```
MEMBERS OF THE LEADING CROWD (COLEMAN, 1964, P.171)

 Response variable: MEM2

 scaled deviance = 1062.7 at cycle  2
           d.f. =    3

 Chi2 probability =  0.      for Chi2 =   1063. with   3. d.f.

         estimate          s.e.      parameter
     1     -0.3654       0.03487      1
     scale parameter taken as   1.000

 scaled deviance = 27.183 (change =  -1036.) at cycle  3
           d.f. =  2      (change =      -1 )

 Chi2 probability =  0.0000 for Chi2 =   27.18 with   2. d.f.

 Chi2 probability =  0.      for Chi2 =   1036. with   1. d.f.

         estimate          s.e.      parameter
     1     -0.1023       0.04248      1
     2        1.249      0.04248      MEM
     scale parameter taken as   1.000

 scaled deviance = 0.022439 (change =   -27.16) at cycle  3
           d.f. = 1          (change =    -1   )

 Chi2 probability =  0.8809 for Chi2 =  0.0224 with   1. d.f.

 Chi2 probability =  0.      for Chi2 =   1063. with   2. d.f.

         estimate          s.e.      parameter
     1     -0.1260       0.04292      1
     2        1.239      0.04266      MEM
     3       0.2183      0.04200      ATT
     scale parameter taken as   1.000
```

We still have one degree of freedom left, which corresponds to the interaction between attitude and membership at the first time point with respect to membership at the second time. This parameter is not necessary for our model. Membership at time point 2 is more probable if one was already a member at point 1 and if one was favourable to the existence of a leading group at that time.

We now take, in our third model, the second attitude as response. We first fit the base model with no explanatory variables,

```
MEMBERS OF THE LEADING CROWD (COLEMAN, 1964, P.171)

 Response variable: ATT2

 scaled deviance = 323.80 at cycle  3
            d.f. =   7

 Chi2 probability =  0.     for Chi2 =    323.8 with    7. d.f.

           estimate          s.e.      parameter
    1       0.2772        0.03464      1
    scale parameter taken as    1.000

  unit  observed     out of      fitted    residual
     1       458        598      340.18       9.729
     2       171        353      200.81      -3.204
     3       184        259      147.34       4.600
     4        85        182      103.53      -2.774
     5       110        159       90.45       3.131
     6        56        143       81.35      -4.280
     7       531        812      461.92       4.895
     8       338        892      507.43     -11.455

MEMBERS OF THE LEADING CROWD (COLEMAN, 1964, P.171)

Binomial Residuals

Score Test Coefficient of Sensitivity

   0.00 I                   S        S        S        S        S
  -4.00 I
  -8.00 I                                                           S
 -12.00 I
 -16.00 I
 -20.00 I
 -24.00 I      S
 -28.00 I
 -32.00 I
 -36.00 I
 -40.00 I
 -44.00 I
 -48.00 I
 -52.00 I
 -56.00 I
 -60.00 I
 -64.00 I                                                                S
 -68.00 I
 -72.00 I
 -76.00 I
---------:---------:---------:---------:---------:---------:---------
       0.00      1.60      3.20      4.80      6.40      8.00      9.
```

```
Residual Plot

   14.40 |
   12.80 |
   11.20 |                                                                   A
    9.60 |                                                                   S
    8.00 |
    6.40 |                                                      A
    4.80 |                                               2      S
    3.20 |                                          2
    1.60 |                                                      Y            Y
    0.00 |                            Y      Y      Y      Y
   -1.60 |          Y        Y
   -3.20 |                            2      2
   -4.80 |                   2
   -6.40 |
   -8.00 |
   -9.60 |
  -11.20 |          S
  -12.80 |          A
  -14.40 |
  -16.00 |
---------:---------:---------:---------:---------:---------:---------
      -2.400    -1.600    -0.800     0.000     0.800     1.600     2.4

Points Y represent 45 line
```

Attitude at the second point in time depends on at least some of the other variables. The score test coefficient of sensitivity indicates that the first and last categories are least well fitted by the model; the number favourable to membership at the second time point is under-estimated for the category, member at both times and favourable the first time, and over-estimated for the category, not member either time and unfavourable the first time.

Next we fit a model with the three main effects, attitude at point one and membership at the two points.

```
MEMBERS OF THE LEADING CROWD (COLEMAN, 1964, P.171)

  scaled deviance = 1.1868 (change =  -322.6) at cycle  3
             d.f. = 4      (change =    -3   )

  Chi2 probability =  0.8805 for Chi2 =   1.187 with   4. d.f.

  Chi2 probability =  0.      for Chi2 =   322.6 with   3. d.f.

          estimate        s.e.       parameter
     1      0.3101      0.03815      1
     2      0.5792      0.03648      ATT1
     3     0.07604      0.04503      MEM1
     4      0.1680      0.04418      MEM2
     scale parameter taken as   1.000

    unit  observed     out of     fitted    residual
       1       458        598     452.36       0.537
       2       171        353     174.29      -0.350
       3       184        259     188.39      -0.612
       4        85        182      82.96       0.303
       5       110        159     109.62       0.066
       6        56        143      58.73      -0.464
       7       531        812     532.63      -0.121
       8       338        892     334.02       0.275
```

```
MEMBERS OF THE LEADING CROWD (COLEMAN, 1964, P.171)
 Binomial Residuals
 Score Test Coefficient of Sensitivity
   0.0000 |                              S
  -0.0600 |
  -0.1200 |                        S                  S
  -0.1800 |                                     S
  -0.2400 |
  -0.3000 |            S
  -0.3600 |
  -0.4200 |
  -0.4800 |                  S
  -0.5400 |
  -0.6000 |
  -0.6600 |
  -0.7200 |
  -0.7800 |                                                 S
  -0.8400 |
  -0.9000 |
  -0.9600 |     S
  -1.0200 |
  -1.0800 |
  -1.1400 |
----------:---------:---------:---------:---------:---------:--------
        0.00      1.60      3.20      4.80      6.40      8.00      9.
 Residual Plot
    1.800 |
    1.600 |                                                Y
    1.400 |
    1.200 |
    1.000 |
    0.800 |                                        Y       A
    0.600 |                                        A       S
    0.400 |                                   2    S
    0.200 |                               Y   S
    0.000 |                               2
   -0.200 |                           3
   -0.400 |                  S    2
   -0.600 |          S       A    A
   -0.800 |          A       Y
   -1.000 |
   -1.200 |
   -1.400 |
   -1.600 |          Y
   -1.800 |
   -2.000 |
----------:---------:---------:---------:---------:---------:--------
       -2.400    -1.600    -0.800    0.000     0.800     1.600     2.4
 Points Y represent 45 line
```

This model fits very well, with four degrees of freedom left; no interaction effects are necessary. The coefficient of sensitivity indicates that the problem with the extremes persists but now both extremes are under-estimated. We shall treat problems with extremes in more detail in Chapter 6.

We may now wonder if any of the main effects might be eliminated. Instead of removing each variable in turn, let us look at the relationships between the parameter estimates and their standard errors. We see that the ratio for membership at point one is considerably less than 2: (0.076/0.045), which is not the case for the other two variables. We shall eliminate it.

```
MEMBERS OF THE LEADING CROWD (COLEMAN, 1964, P.171)

 scaled deviance = 4.0366 (change =  +2.850) at cycle  3
            d.f. = 5      (change =  +1     )
```

chi2 probability = 0.5461 for Chi2 = 4.037 with 5. d.f.

chi2 probability = 0. for Chi2 = 319.8 with 2. d.f.

	estimate	s.e.	parameter
1	0.2969	0.03727	1
2	0.5810	0.03646	ATT1
3	0.2079	0.03737	MEM2

scale parameter taken as 1.000

unit	observed	out of	fitted	residual
1	458	598	447.06	1.030
2	171	353	169.78	0.130
3	184	259	193.63	-1.377
4	85	182	87.53	-0.376
5	110	159	105.18	0.808
6	56	143	54.26	0.301
7	531	812	537.13	-0.455
8	338	892	338.43	-0.030

MEMBERS OF THE LEADING CROWD (COLEMAN, 1964, P.171)

Binomial Residuals

Score Test Coefficient of Sensitivity

```
     0.000 |            S           S            S           S
    -0.120 |                              S
    -0.240 |
    -0.360 |
    -0.480 |
    -0.600 |
    -0.720 |                  S
    -0.840 |
    -0.960 |
    -1.080 |
    -1.200 |                                           S
    -1.320 |
    -1.440 |
    -1.560 |
    -1.680 |
    -1.800 |
    -1.920 |
    -2.040 |
    -2.160 |     S
    -2.280 |
-----------:---------:---------:---------:---------:---------:---------
         0.00      1.60      3.20      4.80      6.40      8.00      9.
```

Residual Plot

```
     1.800 |
     1.600 |                                                Y
     1.400 |                                                A
     1.200 |
     1.000 |                                                S
     0.800 |                                        3
     0.600 |
     0.400 |                                   3
     0.200 |                               3
     0.000 |                           2
    -0.200 |                           Y
    -0.400 |                  S    3
    -0.600 |
    -0.800 |                  2
    -1.000 |
    -1.200 |
    -1.400 |          S
    -1.600 |          2
    -1.800 |
    -2.000 |
-----------:---------:---------:---------:---------:---------:---------
        -2.400    -1.600    -0.800    0.000     0.800     1.600    2.
```

Points Y represent 45 line

The Chi-square is now larger, but not significantly so, and the model is still acceptable. Neither of the other variables can be eliminated.

A look at the coefficient of sensitivity indicates that the extremes no longer pose a problem. Attitude is more favourable at point 2 if one is a member and if attitude was already favourable previously. Previous membership has no significant effect on current attitude when these other two variables are taken into account.

In the analysis of the second wave of the panel, we might usefully have taken a different approach, by using the log linear, instead of the logistic, model and studying all inter-relationships at the same time, instead of in two steps as above. This would be justified by the impossibility of assigning priority to membership or attitude and would remove the arbitrariness noted above. Relationships in the final step above would not change, since parameter values for the log linear and logistic models are identical, but those of the second step (second wave membership as dependent variable) would, since this model does not include second wave attitude.

We may now summarize our findings in a path diagram, as in the previous example:

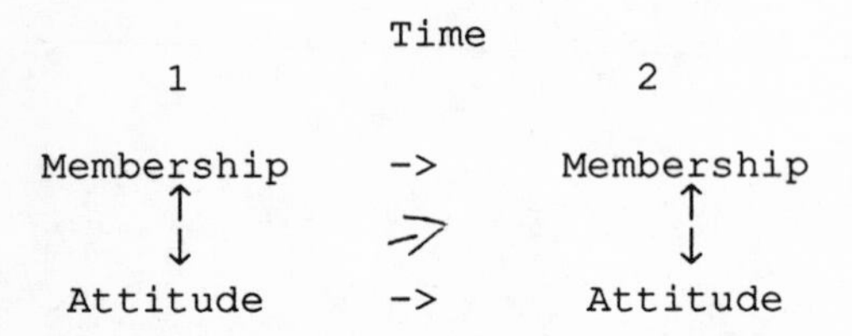

In general, current membership positively affects current attitude, as does previous attitude, but not previous membership. Previous membership and attitude positively affect current membership.

Such path diagrams as the two presented in this chapter are often useful in summarizing certain social relationships. However, their application is very limited, since all variables must be dichotomous and no interactions may be present. Such a situation is rarely the case in the study of any complex social phenomena. In addition, the user must be wary of interpreting such diagrams as demonstrating causality.

4. First Order Markov Chains

When panel observations are available over more than two time periods, it is possible to determine if the same pattern of change occurs in each period. Suppose that individual responses at a given time point depend only on those of the immediately preceding point. This is the hypothesis of a first order Markov chain. Then, the probability of an individual belonging to any given category depends only on his/her category for the immediately preceding time point. We have a square transition matrix of probabilities. If the rows are the categories at the previous time point and the columns are the present categories, then the row probabilities sum to one. This matrix represents the pattern of change; if it is the same over each period, we have stationarity. In this section, we shall test the stationarity of a first order Markov chain, assuming the first order hypothesis. In the next section we shall test the latter hypothesis.

	Party	Republican	Democrat	Undecided
			June	
May	Republican	125	5	16
	Democrat	7	106	15
	Undecided	11	18	142
			July	
June	Republican	124	3	16
	Democrat	6	109	14
	Undecided	22	9	142
			August	
July	Republican	146	2	4
	Democrat	6	111	4
	Undecided	40	36	96
			September	
August	Republican	184	1	7
	Democrat	4	140	5
	Undecided	10	12	82
			October	
September	Republican	192	1	5
	Democrat	2	146	5
	Undecided	11	12	71

Table 2.4 One Step Transitions for Voting Intentions in the 1940 US Presidential Elections, Erie County (Goodman, 1962)

A common application of Markov chains is to voting behaviour. Here, we consider successive monthly expressions of intention to vote in the 1940 US presidential elections for Erie County (Table 2.4). The data consist of a series of five two-way tables, yielding a three-way table over the five time periods.

We have three variables: the voting intention at the beginning of any period (3 categories), the voting intention at the end of any period (3 categories), and the time periods themselves (5 categories). The test for stationarity is a test of independence between time period and intention to vote at the end of the period; this relationship is omitted from the model. Thus, the model will contain the three sets of mean parameters and those for the relationships between intentions at the beginning and end of a period and for those between intentions at the beginning of a period and the period itself. A macro, MPCT, calculates the transition matrix (assuming stationarity) and tests for stationarity.

ONE STEP TRANSITIONS - VOTERS IN ERIE COUNTY, 1940 (GOODMAN, 1962)

First Order Markov Chain

Estimated Stationary Transition Probabilities

```
0.92780  0.01444  0.05776
0.03676  0.90000  0.06324
0.13165  0.12185  0.74650
```

Test for Stationarity

```
scaled deviance = 101.51 at cycle  4
           d.f. =  24

Chi2 probability =  0.0000 for Chi2 =   101.5 with   24. d.f.
```

	estimate	s.e.	parameter
1	4.909	0.08332	1
2	-3.360	0.2308	T1(2)
3	-1.795	0.1484	T1(3)
4	-4.163	0.2909	T2(2)
5	-2.776	0.1488	T2(3)
6	-0.02076	0.1177	TIME(2)
7	0.04027	0.1159	TIME(3)
8	0.2739	0.1098	TIME(4)
9	0.3047	0.1091	TIME(5)
10	7.361	0.3553	T1(2).T2(2)
11	3.319	0.2922	T1(2).T2(3)
12	4.085	0.3267	T1(3).T2(2)
13	4.512	0.1861	T1(3).T2(3)
14	0.02854	0.1715	T1(2).TIME(2)
15	-0.09651	0.1718	T1(2).TIME(3)
16	-0.1220	0.1630	T1(2).TIME(4)
17	-0.1263	0.1620	T1(2).TIME(5)
18	0.03239	0.1596	T1(3).TIME(2)
19	-0.03444	0.1584	T1(3).TIME(3)
20	-0.7712	0.1659	T1(3).TIME(4)
21	-0.9030	0.1685	T1(3).TIME(5)

scale parameter taken as 1.000

ONE STEP TRANSITIONS - VOTERS IN ERIE COUNTY, 1940 (GOODMAN, 1962)

unit	observed	fitted	residual
1	125	135.458	-0.899
2	5	2.108	1.992
3	16	8.433	2.606
4	7	4.706	1.058
5	106	115.200	-0.857
6	15	8.094	2.427
7	11	22.513	-2.426
8	18	20.836	-0.621
9	142	127.651	1.270
10	124	132.675	-0.753
11	3	2.065	0.651
12	16	8.260	2.693
13	6	4.743	0.577
14	109	116.100	-0.659
15	14	8.157	2.046

16	22	22.776	-0.163
17	9	21.080	-2.631
18	142	129.144	1.131
19	146	141.025	0.419
20	2	2.195	-0.132
21	4	8.780	-1.613
22	6	4.449	0.736
23	111	108.900	0.201
24	4	7.651	-1.320
25	40	22.644	3.647
26	36	20.958	3.286
27	96	128.398	-2.859
28	184	178.137	0.439
29	1	2.773	-1.065
30	7	11.090	-1.228
31	4	5.478	-0.631
32	140	134.100	0.509
33	5	9.422	-1.441
34	10	13.692	-0.998
35	12	12.672	-0.189
36	82	77.636	0.495
37	192	183.704	0.612
38	1	2.859	-1.100
39	5	11.437	-1.903
40	2	5.625	-1.528
41	146	137.700	0.707
42	5	9.675	-1.503
43	11	12.375	-0.391
44	12	11.454	0.161
45	71	70.171	0.099

ONE STEP TRANSITIONS - VOTERS IN ERIE COUNTY, 1940 (GOODMAN, 1962)

Poisson Residuals

Score Test Coefficient of Sensitivity

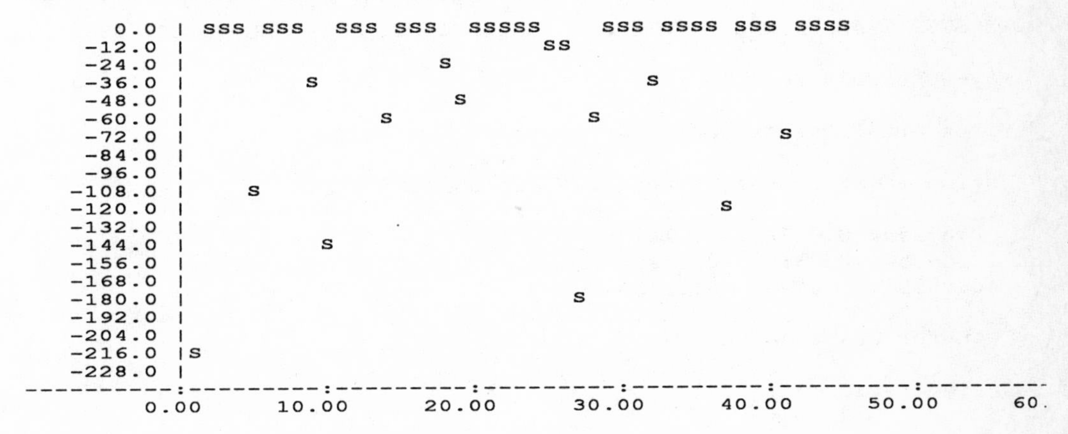

```
Residual Plot

     4.800 |
     4.200 |                                               A    A
     3.600 |                                                    S
     3.000 |                                         AAA A S
     2.400 |                                    AAAAA SS S      Y
     1.800 |                                  AAA   SS   Y Y
     1.200 |                                 AA  SS2YYYY
     0.600 |                               352424YY
     0.000 |                          Y26363Y
    -0.600 |                    YY2Y4244
    -1.200 |               YYYY2SS2S2AA
    -1.800 |           Y Y SSSS AA2A
    -2.400 |      Y    S S    AA
    -3.000 |      S      A AAA
    -3.600 |           A
    -4.200 |
    -4.800 |
    -5.400 |
    -6.000 |
    -6.600 |      A
 ----------:---------:---------:---------:---------:---------:---------:--------
        -3.00     -2.00     -1.00      0.00      1.00      2.00      3
Points Y represent 45 line
```

We see that the hypothesis of stationarity is decisively rejected. The process of changing intentions to vote varies over the months. The score test coefficient of sensitivity indicates that the diagonal elements for the first two periods (1, 5, 9; 10, 14, 18) and those for Democrat to Undecided and Republican to Democrat for the last two periods (29, 33; 38, 42) are poorly estimated. The table of residuals shows that they are all over-estimated except Undecided-Undecided (9; 18). We may conclude that intentions are more stable in the last two periods since the diagonal (no change) is larger then, and that the main differences in changes in intentions between the first and last two periods are the two just mentioned.

One important piece of information is that the Democratic convention was held during the third period. We reconstruct separate tables for the first two time periods and for the last two and apply the macro successively to each:

```
ONE STEP TRANSITIONS - VOTERS IN ERIE COUNTY, 1940 (GOODMAN, 1962)

 May-June-July Period

 First Order Markov Chain

 Estimated Stationary Transition Probabilities

   0.86159  0.02768  0.11073
   0.05058  0.83658  0.11284
   0.09593  0.07849  0.82558

 Test for Stationarity

 scaled deviance = 7.4120 at cycle  3
            d.f. = 6

 Chi2 probability =  0.2838 for Chi2 =   7.412 with   6. d.f.

         estimate        s.e.     parameter
    1       4.835     0.08605     1
    2      -2.967      0.2971     T1(2)
    3      -2.037      0.2015     T1(3)
    4      -3.438      0.3591     T2(2)
```

```
    5      -2.052        0.1878      T2(3)
    6    -0.02076        0.1177      TIME(2)
    7       6.244        0.4588      T1(2).T2(2)
    8       2.854        0.3830      T1(2).T2(3)
    9       3.237        0.4430      T1(3).T2(2)
   10       4.204        0.2628      T1(3).T2(3)
   11     0.02854        0.1715      T1(2).TIME(2)
   12     0.03239        0.1596      T1(3).TIME(2)
    scale parameter taken as   1.000
```

```
ONE STEP TRANSITIONS - VOTERS IN ERIE COUNTY, 1940 (GOODMAN, 1962)

 August-September-October Period

 First Order Markov Chain

 Estimated Stationary Transition Probabilities

   0.964103  0.005120  0.030769
   0.019868  0.947020  0.033113
   0.106061  0.121212  0.772727

 Test for Stationarity

 scaled deviance = 1.4992 at cycle  3
            d.f. = 6

 Chi2 probability =  0.9583 for Chi2 =   1.499 with   6. d.f.

          estimate         s.e.      parameter
    1        5.221      0.07283      1
    2       -4.136       0.4185      T1(2)
    3       -2.820       0.2398      T1(3)
    4       -5.236       0.7081      T2(2)
    5       -3.445       0.2932      T2(3)
    6      0.03077       0.1013      TIME(2)
    7        9.101       0.8193      T1(2).T2(2)
    8        3.956       0.5936      T1(2).T2(3)
    9        5.370       0.7686      T1(3).T2(2)
   10        5.431       0.3743      T1(3).T2(3)
   11    -0.004280       0.1533      T1(2).TIME(2)
   12      -0.1319       0.1747      T1(3).TIME(2)
    scale parameter taken as   1.000
```

In each separate table, stationarity is no longer rejected. We note, as expected, that the diagonal transition probabilities are considerably smaller before the convention than after. In the first table (May-July), 86% of those intending to vote for the Republicans and 84% of those for the Democrats do not change their minds over a month's period, whereas, after the convention (August-October), the percents are 96 and 95 respectively. The coefficients of sensitivity (not shown) no longer exhibit any consistent trend.

Remember, however, that this analysis supposes that intentions at one point in time only depend on intentions one month before. This is the hypothesis of a first order Markov chain.

5. Second Order Markov Chains

In order to test whether a series of observations is a first order Markov chain or one of higher order, we require the details of changes for individuals over successive periods, and not only over one period at a time, as in the preceding section. We can, then, simply test if the categories at the present time point are independent of those two time periods before. If so, the process is of first order. If not, it is *at least* of second order.

We consider again data from the same study as in the previous example. This time, we assume stationarity of second order and collapse the tables (Table 2.5) over the four three-month periods (May-July, June-August, July-September, August-October). This table cannot be obtained from the one in the previous section which only covers two month periods. Note that we could, and should, test for second order stationarity in a way very similar to that of the preceding section.

More concretely, we are assuming that the sequence of changes in voting intentions is identical over any consecutive three month period (stationarity) and we test if current intentions in any month depend only on intentions in the previous month (first order) or depend also on intentions two or more months before (second order or more).

		Time t		
Time t-2	**Time t-1**	Republican	Democrat	Undecided
	Republican	557	6	16
Republican	Democrat	18	0	5
	Undecided	71	1	11
	Republican	3	8	0
Democrat	Democrat	9	435	22
	Undecided	6	63	6
	Republican	17	5	21
Undecided	Democrat	4	10	24
	Undecided	62	54	346

Table 2.5 Two Step Transitions for Voting Intentions in the 1940 US Presidential Elections, Erie County (Goodman, 1962)

We have a three-way table with intentions to vote at times t, t-1, and t-2. We test to see if intentions at time t are independent of those at time t-2:

```
TWO STEP TRANSITIONS - VOTERS IN ERIE COUNTY, 1940 (GOODMAN, 1962)

 scaled deviance = 63.498 at cycle  4
            d.f. = 12

 Chi2 probability =   0.0000 for Chi2 =    63.50 with    12. d.f.

         estimate        s.e.       parameter
  1        6.303      0.04261       1
  2       -3.226       0.2126       T1(2)
  3       -1.942       0.1174       T1(3)
```

4	-7.328	0.3824	T2(2)
5	-4.420	0.1878	T2(3)
6	-4.525	0.3800	T3(2)
7	-3.005	0.1810	T3(3)
8	6.972	0.3717	T1(2).T2(2)
9	3.102	0.3078	T1(2).T2(3)
10	3.862	0.3434	T1(3).T2(2)
11	4.317	0.1980	T1(3).T2(3)
12	7.861	0.4492	T2(2).T3(2)
13	3.447	0.3520	T2(2).T3(3)
14	4.340	0.4135	T2(3).T3(2)
15	4.555	0.2176	T2(3).T3(3)

scale parameter taken as 1.000

unit	observed	fitted	residual
1	557	546.035	0.469
2	6	5.917	0.034
3	16	27.048	-2.124
4	18	21.691	-0.792
5	0	0.235	-0.485
6	5	1.074	3.787
7	71	78.274	-0.822
8	1	0.848	0.165
9	11	3.877	3.617
10	3	0.359	4.410
11	8	10.083	-0.656
12	0	0.558	-0.747
13	9	15.196	-1.589
14	435	427.167	0.379
15	22	23.638	-0.337
16	6	2.446	2.273
17	63	68.750	-0.693
18	6	3.804	1.126
19	17	6.573	4.067
20	5	5.464	-0.199
21	21	30.963	-1.791
22	4	5.808	-0.750
23	10	4.829	2.353
24	24	27.363	-0.643
25	62	70.619	-1.026
26	54	58.707	-0.614
27	346	332.674	0.731

```
TWO STEP TRANSITIONS - VOTERS IN ERIE COUNTY, 1940 (GOODMAN, 1962)
Poisson Residuals
Score Test Coefficient of Sensitivity
      0. |  S    S S  S S SS S    S S  S SS S SS S SS
   -160. |    S S              S     S
   -320. |           S                                S
   -480. |
   -640. |
   -800. |                      S
   -960. |
  -1120. |
  -1280. |
  -1440. |
  -1600. |
  -1760. |
  -1920. |
  -2080. |
  -2240. |
  -2400. |
  -2560. |
  -2720. |
  -2880. | S
  -3040. |
---------:---------:---------:---------:---------:---------:---------
       0.00      6.00      12.00     18.00     24.00     30.00     36.
Residual Plot
   5.400 |
   4.800 |                                          A  A    A
   4.200 |                                       AA    S    S
   3.600 |                                    AA  S S
   3.000 |
   2.400 |                                  AA S S
   1.800 |                                             Y    Y
   1.200 |                                 A  S  YY Y
   0.600 |                                 222YY
   0.000 |                          YYYY333
  -0.600 |                   S 222222222
  -1.200 |                Y 2Y      A
  -1.800 |        Y    2  S       A
  -2.400 |        S             AA
  -3.000 |                     A
  -3.600 |                A AA
  -4.200 |             A
  -4.800 |
  -5.400 |        A
  -6.000 |
---------:---------:---------:---------:---------:---------:---------
      -3.00     -2.00     -1.00      0.00      1.00      2.00      3.
Points Y represent 45 line
```

The hypothesis is rejected. The score test coefficient of sensitivity shows no pattern. As expected with a large Chi-square, the residual plot deviates from the 45 degree line. But, in addition, the residuals do not form a straight line, indicating that the second order model assuming stationarity is poorly chosen. The deviation from 45 degrees results primarily from the poor fit of a second order model, while nonlinearity of the residual plot arises primarily from lack of stationarity.

Given stationarity, present voting intentions depend on more than just those of the preceding time point. They depend, at least, on the two previous points. With sufficient data, higher order Markov hypotheses could be tested in the same manner.

CHAPTER 3

METRIC VARIABLES

1. Time Trends

In Chapter 1, we studied how recall of an event varied in time. There, we had a simple frequency table. We shall now consider a more complex case, where we have a vector of responses at each time point. Our example concerns the attitude towards treatment of criminals by the courts in the USA between 1972 and 1975 (Table 3.1). The five attitudes are: (1) too harshly, (2) not harshly enough, (3) about right, (4) don't know, and (5) no answer. We have the number of responses to each category over a period of five years. Note that this is not a panel, since the same individuals were not asked each time. Thus, we have no indication of how individual opinions change.

	1972	1973	1974	1975
Too Harshly	105	68	42	61
Not Harshly Enough	1066	1092	580	1174
About Right	265	196	72	144
Don't Know	173	138	51	104
No Answer	4	10	8	7

Table 3.1 Changes in Attitudes to Criminals, 1972-1975 (Haberman, 1978, p.128)

We are interested in time trends in opinion, but relative to other opinions. Thus, increase in one category must be studied in relation to a corresponding reduction of other categories. This is exactly what our log linear model does, since marginal totals are fixed with this model.

We first test independence between attitude and year - attitudes do not change over the years:

```
ATTITUDE TO CRIMINALS 1972-1975 - HABERMAN (1978, P.120)

  scaled deviance = 87.051 at cycle  3
            d.f. = 12

  Chi2 probability =  0.     for Chi2 =   87.05 with   12. d.f.
```


The relationship is strongly significant: we must reject independence. Attitudes do change over the years.

We shall now fit a linear trend. To do this, we shall use orthogonal polynomials. These are simply a recoding of the metric variable, here years, as a series of vectors, linear, quadratic, cubic, ... The sum of the product of the elements of any two vectors is zero, the definition of orthogonality. In addition, although not strictly necessary, the sum of the squares of the elements of any such vector is defined as one. The required vectors may be constructed by a macro, found in Appendix III, called ORTH, which creates linear, quadratic, and cubic orthogonal polynomials.

We introduce the linear polynomial for years into our model:

```
ATTITUDE TO CRIMINALS 1972-1975 - HABERMAN (1978, P.120)

 scaled deviance = 13.871 (change =  -73.18) at cycle  3
            d.f. =  8       (change =    -4    )

 Chi2 probability =  0.0846 for Chi2 =   13.87 with   8. d.f.

 Chi2 probability =  0.      for Chi2 =   73.18 with   4. d.f.

           estimate          s.e.      parameter
     1        4.149        0.2427      1
     2       0.1004        0.1600      YEAR(2)
     3      -0.4316        0.3148      YEAR(3)
     4       0.4014        0.4685      YEAR(4)
     5        2.681       0.06406      ATTI(2)
     6       0.8790       0.07399      ATTI(3)
     7       0.5171       0.07847      ATTI(4)
     8       -2.228        0.1963      ATTI(5)
     9       -1.468        0.8234      ATTI(1).YRL
    10      -0.5129        0.7835      ATTI(2).YRL
    11       -1.750        0.7990      ATTI(3).YRL
    12       -1.579        0.8066      ATTI(4).YRL
    13        0.000      aliased      ATTI(5).YRL
     scale parameter taken as   1.000

   unit  observed     fitted    residual
      1       105     98.409       0.664
      2        68     81.121      -1.457
      3        42     35.532       1.085
      4        61     60.938       0.008
      5      1066   1079.176      -0.401
      6      1092   1076.787       0.464
      7       580    570.897       0.381
      8      1174   1185.140      -0.324
      9       265    257.949       0.439
     10       196    200.975      -0.351
     11        72     83.203      -1.228
     12       144    134.873       0.786
     13       173    170.641       0.181
     14       138    137.572       0.036
     15        51     58.934      -1.034
     16       104     98.853       0.518
     17         4      6.825      -1.081
     18        10      7.546       0.893
     19         8      4.433       1.694
     20         7     10.196      -1.001
```

```
ATTITUDE TO CRIMINALS 1972-1975 - HABERMAN (1978, P.120)
Poisson Residuals
Score Test Coefficient of Sensitivity
     0.00 I       S S              S  S    S S  S      S  S
    -2.50 I    S            S
    -5.00 I  S           S       S                S
    -7.50 I                                          S
   -10.00 I                             S
   -12.50 I
   -15.00 I
   -17.50 I
   -20.00 I            S
   -22.50 I
   -25.00 I
   -27.50 I                                                 S
   -30.00 I
   -32.50 I
   -35.00 I
   -37.50 I
   -40.00 I
   -42.50 I                   S
   -45.00 I
   -47.50 I
----------:---------:---------:---------:---------:---------:---------
        0.00      4.00      8.00     12.00     16.00     20.00     24.

Residual Plot
    2.250 I
    2.000 I                                                     Y
    1.750 I                                               A     2
    1.500 I                                           A   Y
    1.250 I                                         A Y
    1.000 I                                 A AA A  Y S   S
    0.750 I                                A      2  S
    0.500 I                              AS S 22
    0.250 I                              SY Y
    0.000 I                           23 Y
   -0.250 I                       S 2 Y
   -0.500 I                      2Y A
   -0.750 I                    Y
   -1.000 I               S 2  S
   -1.250 I           S   Y       A
   -1.500 I     S     Y     A  A A
   -1.750 I               A
   -2.000 I     Y     A
   -2.250 I
   -2.500 I     A
----------:---------:---------:---------:---------:---------:---------
       -2.400    -1.600    -0.800     0.000     0.800     1.600     2.4
Points Y represent 45 line
```

In this case, since the linear effect of the year (YRL) is not included among the main effects, but only as an interaction, GLIM fits the parameters with respect to the *last* category of attitude, set to zero. We see that, in relation to the last attitude (no answer), the other four attitudes all decrease with time, but that the second attitude, that criminals are not treated harshly enough, decreases less than the other three.

A close inspection of the residuals for the independence model (not shown) would already have indicated that the observations for this attitude fitted that model most poorly. This is still evident in the graph giving the score test coefficient of sensitivity for this model.

Instead of attempting to fit a more complex time trend, let us then eliminate this attitude from the model by giving it zero weight in order to see if the remaining attitudes are independent of time:

ATTITUDE TO CRIMINALS 1972-1975 - HABERMAN (1978, P.120)

```
 scaled deviance = 16.575 at cycle  4
            d.f. =  9     from 16 observations

 Chi2 probability =  0.0555 for Chi2 =   16.58 with   9. d.f.

          estimate        s.e.       parameter
     1       4.647     0.06900       1
     2     -0.2834     0.06523       YEAR(2)
     3      -1.151     0.08723       YEAR(3)
     4     -0.5487     0.07066       YEAR(4)
     5       0.000     aliased       ATTI(2)
     6      0.8973     0.07142       ATTI(3)
     7      0.5238     0.07595       ATTI(4)
     8      -2.253      0.1952       ATTI(5)
     scale parameter taken as  1.000

   unit  observed    fitted   residual
      1       105   104.262      0.072
      2        68    78.530     -1.188
      3        42    32.975      1.572
      4        61    60.232      0.099
      5      1066   104.262      0.000
      6      1092    78.530      0.000
      7       580    32.975      0.000
      8      1174    60.232      0.000
      9       265   255.745      0.579
     10       196   192.627      0.243
     11        72    80.885     -0.988
     12       144   147.743     -0.308
     13       173   176.037     -0.229
     14       138   132.591      0.470
     15        51    55.675     -0.627
     16       104   101.696      0.228
     17         4    10.955     -2.101
     18        10     8.251      0.609
     19         8     3.465      2.436
     20         7     6.329      0.267
```

```
ATTITUDE TO CRIMINALS 1972-1975 - HABERMAN (1978, P.120)
Poisson Residuals
Score Test Coefficient of Sensitivity
      0.000 |   S         S                                         S         S
     -0.250 |                                    S       S   S                 S
     -0.500 |                                                    S   S
     -0.750 |
     -1.000 |                                                                 S
     -1.250 |
     -1.500 |           S
     -1.750 |      S
     -2.000 |                                  S
     -2.250 |                                          S
     -2.500 |
     -2.750 |
     -3.000 |
     -3.250 |
     -3.500 |
     -3.750 |
     -4.000 |
     -4.250 |
     -4.500 |
     -4.750 |                                                            S
-----------:----------:----------:----------:----------:----------:---------
         0.00       4.00       8.00      12.00      16.00      20.00      24.
```

Residual Plot

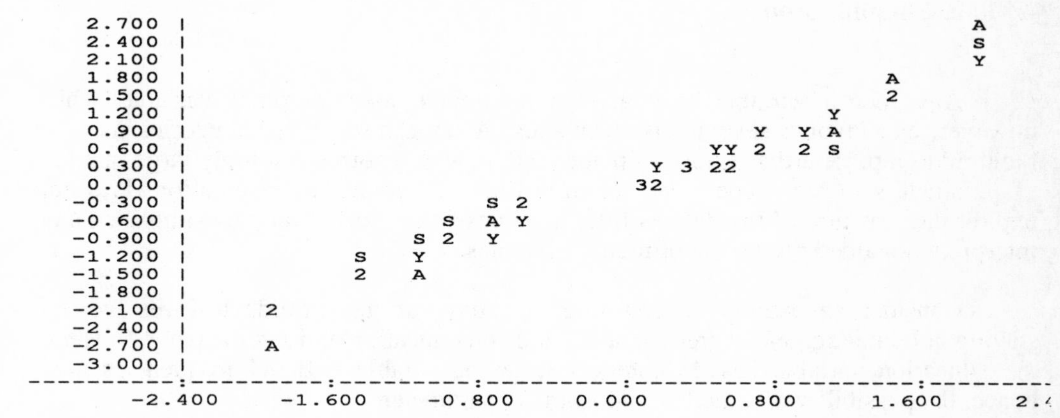

Points Y represent 45 line

The model is just non-significant so that all attitudes except "not harshly enough" do not appear to change with time. Inspection of the residuals for this model indicates, however, that the "no answer" response may also be varying with time. Elimination of this category significantly improves the fit and leaves a very acceptable model:

```
ATTITUDE TO CRIMINALS 1972-1975 - HABERMAN (1978, P.120)

  scaled deviance = 5.7700 at cycle  3
             d.f. = 6        from 12 observations

  Chi2 probability =  0.4502 for Chi2 =   5.770 with   6. d.f.

  Chi2 probability =  0.0129 for Chi2 =   10.81 with   3. d.f.

            estimate         s.e.      parameter
      1        4.660      0.06899      1
      2      -0.3007      0.06580      YEAR(2)
      3       -1.191      0.08889      YEAR(3)
      4      -0.5638      0.07126      YEAR(4)
```

```
5        0.000      aliased      ATTI(2)
6       0.8973      0.07142      ATTI(3)
7       0.5238      0.07595      ATTI(4)
8        0.000      aliased      ATTI(5)
scale parameter taken as   1.000
```

The categories (1) too harshly, (3) about right, and (4) don't know apparently do not vary, with respect to each other, over the four years. Inspection of the residuals and plots (not shown) indicates no systematic patterns.

Our two approaches are complementary. Introduction of orthogonal polynomials indicates how certain attitudes change, while elimination of these attitude categories allows a test isolating those attitudes which have not changed among themselves. We conclude that the attitudes "too harshly", "about right", and "don't know" remain relatively stable in relation to each other and are all losing ground to the "not harshly enough" attitude.

2. Model Simplification

Any metric variable may always be treated as a nominal variable. This, however, can involve several disadvantages. A large number of categories may be required to represent the relationship adequately, with a correspondingly large number of parameters in the model. At the same time, a nominal representation does not exploit the structure of the data as fully as is possible. Models may be simplified and interpretation aided by the use of metric variables.

Consider an example concerning the study of the attitude towards women staying at home (agree/disagree) as it depends on education and sex (Table 3.2). Here, the education variable has 21 categories, giving a table with 84 frequencies and, hence, the possibility of a model with as many parameters.

With a binary response, we use the binomial distribution. However, attempting to fit the data as they stand immediately poses a problem with GLIM: there are no women at education level 2. GLIM gives an error that the binomial denominator cannot be zero. This may be circumvented by giving an arbitrary value to this denominator (for women with education level 2) and specifying a weight of zero, so that this observation is ignored (see the program in Appendix II).

The base model, with only a general mean, is highly significant:

```
ATTITUDE TO WOMEN STAYING AT HOME - HABERMAN (1979, P.312)

 scaled deviance = 451.72 at cycle  3
            d.f. =  40    from 41 observations

 Chi2 probability =  0.      for Chi2 =    451.7 with    40. d.f.

          estimate         s.e.       parameter
    1     -0.5959       0.03899       1
    scale parameter taken as   1.000
```

Attitudes to women staying at home depend either on sex or on education or on both.

Agree	Disagree	Sex	Education
4	2	M	0
4	2	F	0
2	0	M	1
1	0	F	1
4	0	M	2
0	0	F	2
6	3	M	3
6	1	F	3
5	5	M	4
10	0	F	4
13	7	M	5
14	7	F	5
25	9	M	6
17	5	F	6
27	15	M	7
26	16	F	7
75	49	M	8
91	36	F	8
29	29	M	9
30	35	F	9
32	45	M	10
55	67	F	10
36	59	M	11
50	62	F	11
115	245	M	12
190	403	F	12
31	70	M	13
17	92	F	13
28	79	M	14
18	81	F	14
9	23	M	15
7	34	F	15
15	110	M	16
13	115	F	16
3	29	M	17
3	28	F	17
1	28	M	18
0	21	F	18
2	13	M	19
1	2	F	19
3	20	M	20
2	4	F	20

Table 3.2 Attitude to Women Staying at Home with Respect to Sex and Educational Level (Haberman, 1979, p.312)

A model with only sex as the explanatory variable must also be rejected. At this point, differences by sex appear not to be important, since the fit is not improved over the independence model:

```
ATTITUDE TO WOMEN STAYING AT HOME - HABERMAN (1979, P.312)

 scaled deviance = 451.71 (change =  -0.011) at cycle  3
            d.f. =  39    (change =  -1     ) from 41 observations

 Chi2 probability =  0.      for Chi2 =   451.7 with   39. d.f.

 Chi2 probability =  0.9150 for Chi2 =  0.0114 with   1. d.f.

          estimate         s.e.      parameter
      1     -0.5955      0.03915      1
      2    0.004181      0.03915      SEX
      scale parameter taken as  1.000
```

Introduction of a nominal education variable in place of sex explains a lot of the variability and leaves a non-significant lack of fit, but we have a very complex model:

```
ATTITUDE TO WOMEN STAYING AT HOME - HABERMAN (1979, P.312)

 scaled deviance = 27.657 (change =  -424.1) at cycle  9
            d.f. = 20     (change =    -19  ) from 41 observations

 Chi2 probability =  0.1174 for Chi2 =   27.66 with   20. d.f.

 Chi2 probability =  0.      for Chi2 =   424.1 with   19. d.f.

          estimate          s.e.      parameter
      1     0.6931       0.6124      1
      2      8.514        34.97      EDUC(2)
      3      8.866        36.11      EDUC(3)
      4     0.4055       0.8416      EDUC(4)
      5     0.4055       0.8010      EDUC(5)
      6   -0.03637       0.6953      EDUC(6)
      7     0.4055       0.6857      EDUC(7)
      8    -0.1568       0.6528      EDUC(8)
      9   -0.02381       0.6267      EDUC(9)
     10    -0.7745       0.6384      EDUC(10)
     11    -0.9457       0.6288      EDUC(11)
     12     -1.035       0.6284      EDUC(12)
     13     -1.447       0.6163      EDUC(13)
     14     -1.910       0.6340      EDUC(14)
     15     -1.940       0.6348      EDUC(15)
     16     -1.964       0.6746      EDUC(16)
     17     -2.777       0.6443      EDUC(17)
     18     -2.944       0.7478      EDUC(18)
     19     -4.585        1.181      EDUC(19)
     20     -2.303       0.8803      EDUC(20)
     21     -2.262       0.7853      EDUC(21)
      scale parameter taken as  1.000
```

From the parameter estimates, we see that, in general, it is more probable for those with lower education levels to be favourable to women staying at home. In the binomial model, we are studying the relation agree/disagree. The larger parameter estimates at low education levels than at high indicate more chance of agreeing than disagreeing at these low levels as compared to the higher education levels.

We shall now try the linear trend variable for education completed, using the macro ORTH:

```
ATTITUDE TO WOMEN STAYING AT HOME - HABERMAN (1979, P.312)

 scaled deviance = 64.025 (change =  +36.37) at cycle  4
            d.f. = 39     (change =  +19    ) from 41 observations

 Chi2 probability =  0.0070 for Chi2 =   64.03 with   39. d.f.

         estimate          s.e.      parameter
    1     -0.2032       0.04606      1
    2      -10.62        0.6048      EDL
    scale parameter taken as  1.000
```

Although the model is very significantly improved over the base model and the model with only sex, the Chi-square for lack of fit with respect to the saturated model is still very significant. From the negative parameter estimate, we now see more clearly that agreeing that women should stay at home decreases with increasing education: the ratio of the number of people agreeing to disagreeing decreases as education increases.

We next put sex back into the model and add an interaction between sex and the linear effect of education. Note that, since no factor variable is involved, this must be calculated as a new variable before introducing it into the $Fit.

```
ATTITUDE TO WOMEN STAYING AT HOME - HABERMAN (1979, P.312)

 scaled deviance = 57.103 (change = -6.9225) at cycle  4
            d.f. = 37     (change = -2      ) from 41 observations

 Chi2 probability =  0.0185 for Chi2 =   57.10 with   37. d.f.

 Chi2 probability =  0.0314 for Chi2 =   6.922 with   2. d.f.

         estimate          s.e.      parameter
    1     -0.1966       0.04634      1
    2      -10.78        0.6101      EDL
    3    -0.04545       0.04634      SEX
    4       1.597        0.6101      ESL
    scale parameter taken as  1.000

ATTITUDE TO WOMEN STAYING AT HOME - HABERMAN (1979, P.312)

   unit  observed     out of     fitted    residual
      1         4          6      5.344      -1.759
      2         4          6      5.716      -3.301
      3         2          2      1.732       0.557
      4         1          1      0.936       0.261
      5         4          4      3.345       0.885
      6         0          1      0.915       0.000
      7         6          9      7.214      -1.015
      8         6          7      6.206      -0.246
      9         5         10      7.617      -1.943
     10        10         10      8.509       1.324
     11        13         20     14.334      -0.662
     12        14         21     16.931      -1.619
```

13	25	34	22.673	0.847
14	17	22	16.549	0.223
15	27	42	25.747	0.397
16	26	42	28.935	-0.978
17	75	124	68.976	1.089
18	91	127	78.443	2.293
19	29	58	28.883	0.031
20	30	65	35.163	-1.285
21	32	77	33.863	-0.428
22	55	122	56.398	-0.254
23	36	95	36.401	-0.085
24	50	112	43.168	1.327
25	115	360	118.645	-0.409
26	190	593	186.137	0.342
27	31	101	28.286	0.601
28	17	109	27.275	-2.272
29	28	107	25.186	0.641
30	18	99	19.383	-0.350
31	9	32	6.268	1.217
32	7	41	6.183	0.356
33	15	125	20.202	-1.264
34	13	128	14.681	-0.466
35	3	32	4.235	-0.644
36	3	31	2.677	0.207
37	1	29	3.124	-1.272
38	0	21	1.354	-1.203
39	2	15	1.308	0.633
40	1	3	0.144	2.315
41	3	23	1.617	1.129
42	2	6	0.212	3.950

ATTITUDE TO WOMEN STAYING AT HOME - HABERMAN (1979, P.312)

Binomial Residuals

Score Test Coefficient of Sensitivity

```
    0.000 |  SSS SS SS SS    S SSS   S SSSS SSS SSSS
   -0.160 |SS      S  S  S   S   SSS          S    S
   -0.320 |          S
   -0.480 |               S
   -0.640 |                          S    S
   -0.800 |
   -0.960 |
   -1.120 |
   -1.280 |
   -1.440 |
   -1.600 |
   -1.760 |
   -1.920 |
   -2.080 |
   -2.240 |
   -2.400 |
   -2.560 |
   -2.720 |                 S
   -2.880 |
   -3.040 |
----------:---------:---------:---------:---------:---------:---------
        0.00      10.00     20.00     30.00     40.00     50.00     60.
```

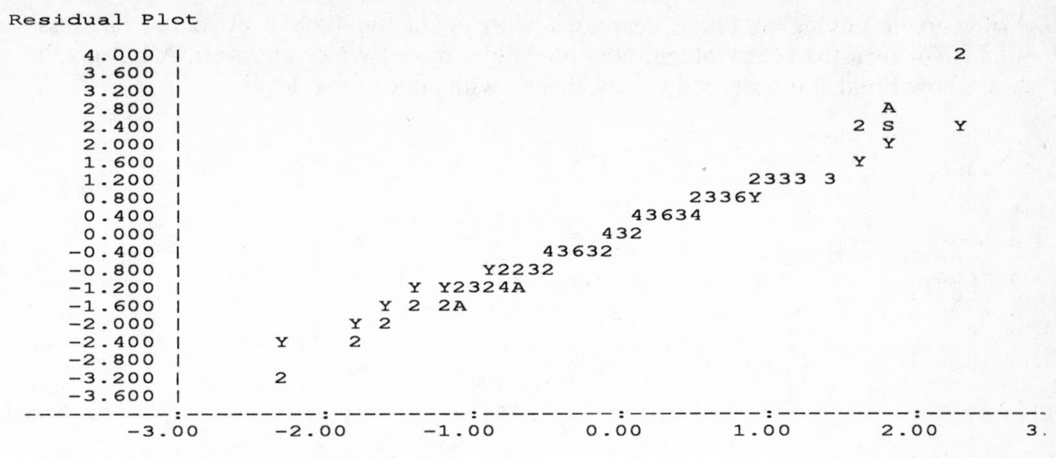

The model is significantly better but still not sufficiently good. We may note that agreement with women staying at home is about the same, on average, for the two sexes (-0.045), that it decreases with education (-10.78), but that it decreases less quickly for men (-10.78 + 1.597 = -9.68) than for women (-10.78 - 1.597 = -12.38): the interaction effect.

If we add the quadratic main effect and its interaction with sex, the Chi-square virtually does not change at all. If we want a simple model, we seem to be left with one which does not fit the data sufficiently well. However, if we look at the residuals for our model, we see that observation 18 (female education level 8) fits the data much less well than the others. This may be what is known as an outlier. We may have an anomaly in the data which should be checked with the original coding sheets and questionnaires. Unfortunately, this is not possible in the secondary analysis of data which we carry out here.

Study of the list of residuals for our linear model also shows that the large residuals are primarily for education levels less than six. This is not obvious from the score test coefficient of sensitivity, because of the scale imposed by the large value for observation 18. We may try eliminating these lower levels from our model:

```
ATTITUDE TO WOMEN STAYING AT HOME - HABERMAN (1979, P.312)

 scaled deviance = 36.018 at cycle  3
            d.f. = 26     from 30 observations

 Chi2 probability =  0.0911 for Chi2 =    36.02 with    26. d.f.

         estimate          s.e.       parameter
     1    -0.1489       0.04959       1
     2   -0.03546       0.04959       SEX
     3     -11.63         0.6891      EDL
     4      1.508         0.6891      ESL
     scale parameter taken as  1.000
```

We now have a simple model which fits very well. The parameter estimates have changed very little from the previous model. Our conclusions above, that agreement with women staying at home decreases with education (above 5 years), but less quickly for men than for women, now holds in a model which fits well. A last graph shows how fitted and observed values change with educational level.

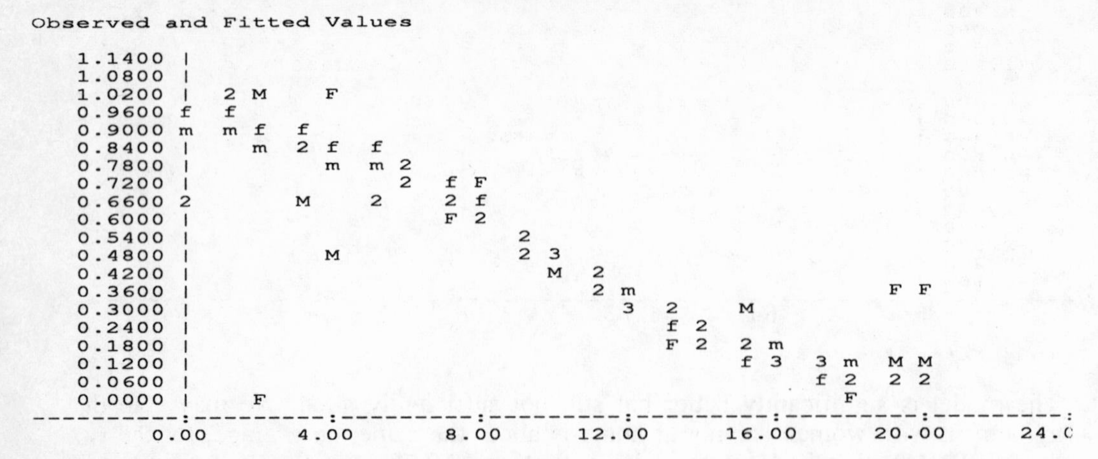

This graph presents two curves, one for men (m) and another for women (f) with the corresponding observed values (M and F) around them. Such curves of fitted values are known as logistic curves. Notice how they flatten off on top at one without quite reaching it and the same at the bottom before reaching zero. The curve for men starts off lower than that for women and ends higher, i.e. is flatter. The contrast between low and high education is greater for women than for men, as was already indicated by the parameter values.

CHAPTER 4

ORDINAL VARIABLES

1. The Log-Multiplicative Model I

In the analysis of categorical data, one commonly encountered type is the ordinal variable. The categories are known to have an order but knowledge of the scale is insufficient to consider them to form a metric. Thus, ordinal variables lie in between nominal and metric variables. In this chapter, we shall employ several approaches to such variables. The first, and perhaps most obvious, is to estimate a scale upon which the values of the variable lie. However, such a scale is never unique. It must always be calculated in relation to one or more other variables. In this way, the choice of criterion variables determines the resulting scale, which varies with the choice. The scale is estimated by successive approximations and then, finally, fitted as if it were a metric variable. In this section, we consider the case of a table with one ordinal variable and one or more nominal variables. In the next section, we apply the same principle to tables with two ordinal variables.

When the scale for a nominal variable is estimated, the model is called log-multiplicative, since it is no longer linear in the unknown parameters, but multiplicative on the log scale for these parameters. Two unknown parameters are multiplied together:

$$\log(F_{ik}) = \mu + \theta_i + \phi_k + \alpha_i \upsilon \qquad (4.1)$$

Both α_i, indexing the nominal variable(s) and υ, the scale to be estimated, are unknown parameters. Since the model is not linear in these parameters, estimation cannot be done directly with the existing GLIM algorithm, but must proceed iteratively, by successive approximations. A macro, L1OV, provided in Appendix III, performs the required calculations and prints out the scale.

We shall apply the model to data on criminal cases in North Carolina (Table 4.1). The ordinal variable is the outcome of the case and the three explanatory variables are race, type of offence, and county. In order to be able to apply the macro, these three variables must combined as one complex variable with 20 categories. This is equivalent to including all possible interactions among these variables, in relation to the ordinal variable, in the model. After we have obtained the scale, we shall verify if all such interactions are necessary, or if some may be eliminated. Thus, our nominal scale will be constructed in relation to this set of three variables.

Offence	Race	County	Outcome: Not Prosecuted	Guilty	Not Guilty
Drinking	Black	Durham	33	8	4
Violence	Black	Durham	10	10	3
Property	Black	Durham	9	8	2
Traffic	Black	Durham	4	2	1
Speeding	Black	Durham	32	3	0
Drinking	Black	Orange	5	10	1
Violence	Black	Orange	5	5	5
Property	Black	Orange	11	5	3
Traffic	Black	Orange	12	6	1
Speeding	Black	Orange	20	3	2
Drinking	White	Durham	53	2	2
Violence	White	Durham	7	8	1
Property	White	Durham	10	5	2
Traffic	White	Durham	16	3	2
Speeding	White	Durham	87	5	3
Drinking	White	Orange	14	2	0
Violence	White	Orange	1	5	7
Property	White	Orange	5	4	0
Traffic	White	Orange	13	13	1
Speeding	White	Orange	98	16	7

Table 4.1 Outcomes of Criminal Case in North Carolina, Classified by Type of Offence, County, and Race (Upton, 1978, p.104)

The macro first fits the usual model for independence. This is independence between the ordinal variable, outcome of the case, taken as a nominal variable, and the three explanatory variables:

```
CRIMINAL CASES IN N. CAROLINA, OFFENCE, COUNTY, RACE (UPTON, 1978, P.

 Independence Model

 scaled deviance = 156.23 at cycle  5
           d.f.  = 38

 Chi2 probability =  0.    for Chi2 =   156.2 with   38. d.f.

          estimate       s.e.      parameter
     1     3.483       0.1511      1
     2    -0.6712      0.2563      IND(2)
     3    -0.8622      0.2736      IND(3)
     4    -1.861       0.4063      IND(4)
     5    -0.2513      0.2254      IND(5)
     6    -1.034       0.2911      IND(6)
     7    -1.099       0.2981      IND(7)
     8    -0.8622      0.2736      IND(8)
     9    -0.8622      0.2736      IND(9)
    10    -0.5878      0.2494      IND(10)
    11     0.2364      0.1994      IND(11)
    12    -1.034       0.2911      IND(12)
    13    -0.9734      0.2847      IND(13)
    14    -0.7621      0.2643      IND(14)
    15     0.7472      0.1810      IND(15)
    16    -1.034       0.2911      IND(16)
    17    -1.242       0.3147      IND(17)
```

```
 18        -1.609          0.3651       IND(18)
 19       -0.5108          0.2434       IND(19)
 20        0.9891          0.1746       IND(20)
 21        -1.286          0.1019       OUT(2)
 22        -2.248          0.1534       OUT(3)
  scale parameter taken as   1.000
```

We see that this independence is decisively rejected. The outcomes of criminal cases in North Carolina depend on one or more of the variables, county, race, and type of offence.

The macro then continues automatically and treats the variable of interest as metric and linear. This is equivalent to assuming that the ordinal scale has equal intervals for its categories:

```
Linear Effects Model

scaled deviance = 40.201 at cycle  4
           d.f. = 19

Chi2 probability =   0.0031 for Chi2 =    40.20 with   19. d.f.

          estimate            s.e.       parameter
  1          3.816          0.2849       1
  2        -0.3148          0.3150       IND(2)
  3        -0.5445          0.3383       IND(3)
  4         -1.588          0.4928       IND(4)
  5         -1.375          0.5906       IND(5)
  6        -0.6520          0.3431       IND(6)
  7        -0.6660          0.3399       IND(7)
  8        -0.5837          0.3455       IND(8)
  9        -0.7562          0.3803       IND(9)
 10        -0.7508          0.4020       IND(10)
 11        -0.7071          0.4416       IND(11)
 12        -0.7208          0.3569       IND(12)
 13        -0.7392          0.3665       IND(13)
 14        -0.8048          0.3997       IND(14)
 15        -0.1142          0.3585       IND(15)
 16         -1.830          0.6942       IND(16)
 17        -0.9988          0.3987       IND(17)
 18         -1.471          0.4904       IND(18)
 19        -0.2523          0.3166       IND(19)
 20         0.7380          0.2744       IND(20)
 21         -1.466          0.1903       OUT(2)
 22         -3.161          0.3873       OUT(3)
 23         0.1791          0.1550       IND(1).ZZ4_
 24         0.5440          0.1662       IND(2).ZZ4_
 25         0.4867          0.1811       IND(3).ZZ4_
 26         0.4297          0.2801       IND(4).ZZ4_
 27        -0.5150          0.2988       IND(5).ZZ4_
 28         0.5904          0.1871       IND(6).ZZ4_
 29         0.7902          0.1881       IND(7).ZZ4_
 30         0.4370          0.1846       IND(8).ZZ4_
 31         0.2656          0.2018       IND(9).ZZ4_
 32        0.05981          0.2084       IND(10).ZZ4_
 33        -0.4154          0.2199       IND(11).ZZ4_
 34         0.4806          0.1936       IND(12).ZZ4_
 35         0.3874          0.1970       IND(13).ZZ4_
```

```
36        0.1465        0.2098      IND(14).ZZ4_
37       -0.3693        0.1740      IND(15).ZZ4_
38       -0.3323        0.3594      IND(16).ZZ4_
39         1.184        0.2241      IND(17).ZZ4_
40        0.2938        0.2717      IND(18).ZZ4_
41        0.4140        0.1642      IND(19).ZZ4_
42         0.000       aliased      IND(20).ZZ4_
 scale parameter taken as    1.000
```

The macro has created a new variable, called ZZ4_, identical in value to the variable outcome (OUT), but defined as a metric variable instead of as a factor variable. Thus, both OUT and ZZ4_ are included in the model, the first as a main effect and the second in the interaction.

The fit is improved, showing that outcome is related to race, county, and type of offence, but the lack of fit is still significant. The ordinal scale appears not to be equally spaced. As for one example in the previous chapter, comparisons are with respect to the last category of the complex independent variable (IND). We see that every fifth parameter estimate is negative (-0.515, -0.369) or about zero (0.0598, 0.000). This indicates that the slope for speeding offences is smaller than that of the other offences: this offence is less often prosecuted that the other four. As well, the slopes for drinking for Whites are also negative (-0.415, -0.332).

Finally, the macro fits the log multiplicative model:

```
Log Multiplicative Model

scaled deviance =     23.85 at cycle    5.
           d.f. =     18.

Scale for ordinal variable
  0.        0.8966    1.000

Chi2 probability =   0.2015 for Chi2 =     23.85 with    19. d.f.

          estimate          s.e.      parameter
   1         3.494        0.1743      1
   2        -1.183        0.3595      IND(2)
   3        -1.286        0.3742      IND(3)
   4        -2.112        0.5302      IND(4)
   5      -0.02536        0.2481      IND(5)
   6        -1.834        0.4687      IND(6)
   7        -1.930        0.4885      IND(7)
   8        -1.104        0.3491      IND(8)
   9        -1.000        0.3360      IND(9)
  10       -0.5019        0.2838      IND(10)
  11        0.4745        0.2220      IND(11)
  12        -1.523        0.4114      IND(12)
  13        -1.192        0.3609      IND(13)
  14       -0.7260        0.3052      IND(14)
  15        0.9709        0.2046      IND(15)
  16       -0.8506        0.3186      IND(16)
  17        -3.862         1.188      IND(17)
  18        -1.860        0.4745      IND(18)
  19       -0.9034        0.3243      IND(19)
  20         1.090        0.2014      IND(20)
```

21	-1.731	0.2294	OUT(2)
22	-2.824	0.2796	OUT(3)
23	0.4784	0.4413	IND(1).ZZ1_
24	1.828	0.5170	IND(2).ZZ1_
25	1.653	0.5544	IND(3).ZZ1_
26	1.262	0.8603	IND(4).ZZ1_
27	-1.035	0.7087	IND(5).ZZ1_
28	2.332	0.6243	IND(6).ZZ1_
29	2.381	0.6457	IND(7).ZZ1_
30	1.237	0.5596	IND(8).ZZ1_
31	0.9535	0.5726	IND(9).ZZ1_
32	0.08274	0.5931	IND(10).ZZ1_
33	-1.204	0.6089	IND(11).ZZ1_
34	1.783	0.5963	IND(12).ZZ1_
35	1.177	0.5875	IND(13).ZZ1_
36	0.3249	0.6052	IND(14).ZZ1_
37	-1.006	0.4701	IND(15).ZZ1_
38	-0.5802	0.8660	IND(16).ZZ1_
39	4.658	1.318	IND(17).ZZ1_
40	1.259	0.7680	IND(18).ZZ1_
41	1.588	0.4844	IND(19).ZZ1_
42	0.000	aliased	IND(20).ZZ1_

scale parameter taken as 1.000

CRIMINAL CASES IN N. CAROLINA, OFFENCE, COUNTY, RACE (UPTON, 1978, P.10

Poisson Residuals

Score Test Coefficient of Sensitivity

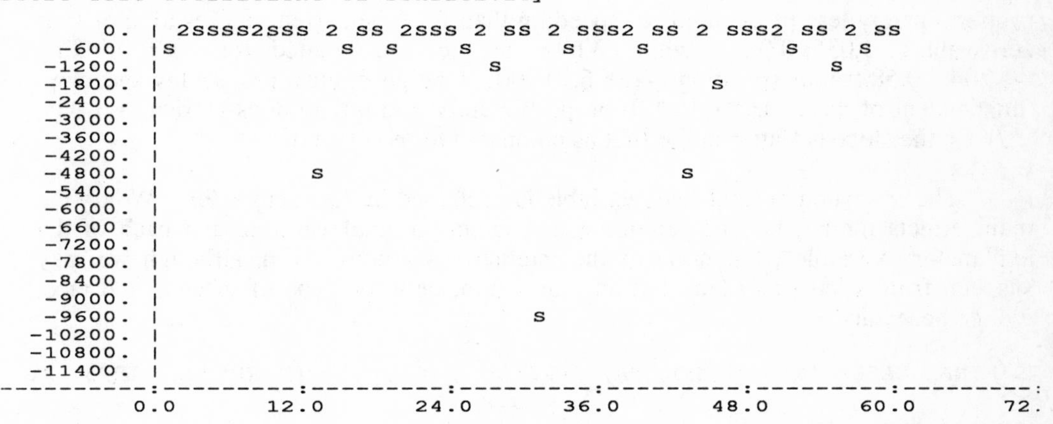

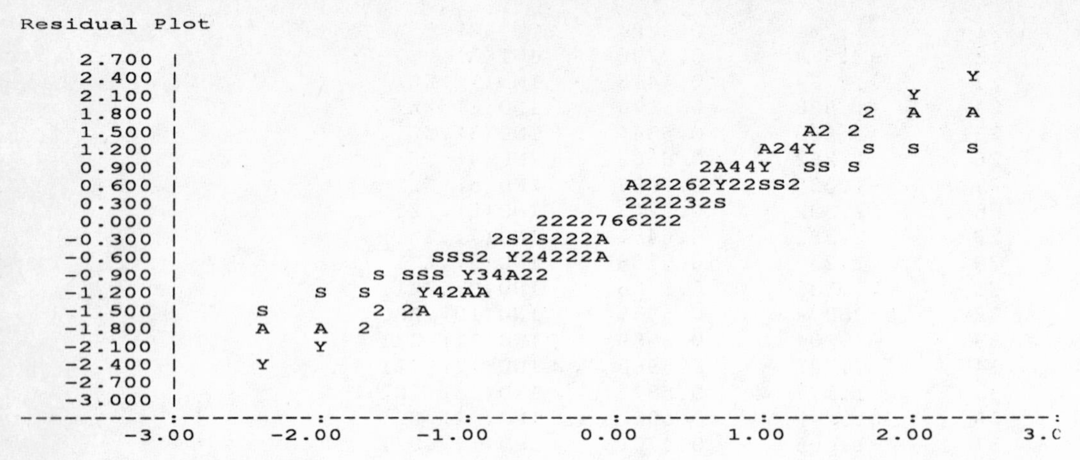

We see that the scale places the two prosecution outcomes close together, separated from the no prosecution outcome. Now, the fit is good, perhaps too good, since a large number of parameters have been included to represent all the interactions. In fact, the residual plot has a slope of less than 45 degrees (whereas poorly fitting models have plots with slopes greater than 45) and the score test indicates the categories with large frequencies as fitting poorly.

The pattern of values for the estimates remains the same. Blacks are proportionately less prosecuted for speeding than for other offences (negative or near zero values: -1.035, 0.083), while Whites are less prosecuted for both drinking (-1.204, -0.580) and speeding (-1.006, 0.000). Orange county prosecutes speeding, independent of race, (0.083, 0.000) proportionately more than does Durham (-1.035, -1.006); the slope is flatter in the first as compared to the second.

The estimated ordinal scale variable is contained in the vector ZZ1_. We try the main effects model, i.e. the relation between this ordinal variable and each of the explanatory variables, but none of the interactions among them, although we may suspect, from what preceded, that an interaction, between type of offence and race, will be necessary.

```
CRIMINAL CASES IN N. CAROLINA, OFFENCE, COUNTY, RACE (UPTON, 1978, P.

 scaled deviance = 50.895 at cycle  4
           d.f. = 32

 Chi2 probability =  0.0183 for Chi2 =   50.89 with   32. d.f.

          estimate        s.e.      parameter
    1        3.575      0.1575      1
    2       -2.303      0.2687      OUT(2)
    3       -3.441      0.3216      OUT(3)
    4       -1.415      0.3110      IND(2)
    5       -1.133      0.2963      IND(3)
    6       -2.009      0.4175      IND(4)
    7      -0.1437      0.2305      IND(5)
    8       -1.263      0.3009      IND(6)
    9       -2.419      0.3888      IND(7)
```

```
  10       -1.534       0.3282      IND(8)
  11       -1.343       0.3108      IND(9)
  12      -0.6154       0.2585      IND(10)
  13       0.2962       0.2033      IND(11)
  14       -1.585       0.3385      IND(12)
  15       -1.127       0.3063      IND(13)
  16      -0.8172       0.2785      IND(14)
  17       0.8881       0.1886      IND(15)
  18       -1.153       0.3016      IND(16)
  19       -2.310       0.4015      IND(17)
  20       -2.098       0.4067      IND(18)
  21      -0.8363       0.2772      IND(19)
  22        1.028       0.1869      IND(20)
  23       0.7374       0.3132      ZZ1_.OFF(1)
  24        2.733       0.3539      ZZ1_.OFF(2)
  25        1.732       0.3488      ZZ1_.OFF(3)
  26        1.357       0.3251      ZZ1_.OFF(4)
  27        0.000      aliased      ZZ1_.OFF(5)
  28       0.8749       0.2257      ZZ1_.COUN(2)
  29      -0.3574       0.2266      ZZ1_.RACE(2)
   scale parameter taken as   1.000
```

The fit is not sufficiently good. We gain 14 d.f., but have eliminated too many parameters. We see that, on average, speeding is less often prosecuted than the other offences, that Orange county prosecutes more, on average, than Durham, and that Whites are less often prosecuted than Blacks.

The different interactions might now be tried, but we quickly discover that it is sufficient to add the interaction between type of offence and race in relation to outcome in order to obtain a satisfactory model:

```
CRIMINAL CASES IN N. CAROLINA, OFFENCE, COUNTY, RACE (UPTON, 1978, P.

 scaled deviance = 35.226 (change =   -15.67) at cycle  4
            d.f. = 28     (change =    -4    )

 Chi2 probability =  0.1631 for Chi2 =   35.23 with   28. d.f.

         estimate          s.e.      parameter
   1        3.416       0.1736      1
   2       -2.539       0.3926      OUT(2)
   3       -3.712       0.4540      OUT(3)
   4       -1.045       0.3257      IND(2)
   5      -0.9386       0.3130      IND(3)
   6       -1.758       0.4283      IND(4)
   7      0.04069       0.2443      IND(5)
   8       -1.376       0.3119      IND(6)
   9       -1.990       0.4131      IND(7)
  10       -1.324       0.3533      IND(8)
  11       -1.034       0.3388      IND(9)
  12      -0.4069       0.2764      IND(10)
  13       0.5572       0.2198      IND(11)
  14       -1.724       0.4134      IND(12)
  15       -1.013       0.3294      IND(13)
  16      -0.7063       0.2914      IND(14)
  17        1.041       0.2015      IND(15)
  18      -0.7942       0.3120      IND(16)
```

```
19       -2.540      0.4933     IND(17)
20       -2.015      0.4461     IND(18)
21      -0.7614      0.3009     IND(19)
22        1.174      0.2008     IND(20)
23        1.659      0.5106     ZZ1_.OFF(1)
24        2.607      0.5546     ZZ1_.OFF(2)
25        1.896      0.5481     ZZ1_.OFF(3)
26        1.270      0.6083     ZZ1_.OFF(4)
27        0.000     aliased     ZZ1_.OFF(5)
28       0.8820      0.2309     ZZ1_.COUN(2)
29       -2.050      0.5463     ZZ1_.RACE(2)
30        2.619      0.7988     ZZ1_.OFF(2).RACE(2)
31        1.941      0.7868     ZZ1_.OFF(3).RACE(2)
32        2.240      0.7758     ZZ1_.OFF(4).RACE(2)
33        2.030      0.7181     ZZ1_.OFF(5).RACE(2)
  scale parameter taken as   1.000
```

Again, we see that offences of violence (2.607) are more often at the high end of the scale (i.e. most prosecuted) and speeding (0.000) at the low end, that Orange County prosecutes more (0.882), and that Whites are prosecuted less (-2.050). The interaction now shows that Whites are prosecuted proportionately less than Blacks for drunkenness; for Whites (race 2), the four interaction parameters are all positive as compared to zero for the first category, drinking.

The score test and residual plots (not shown) are now as expected for an acceptable model: no obvious pattern in the score test coefficient of sensitivity and the residual plot lying at 45 degrees.

2. The Log-Multiplicative Model II

When a table contains two ordinal variables, a scale may be estimated for each of them in relation to the other. We still have a log-multiplicative model:

$$\log(F_{ik}) = \mu + \theta_i + \phi_k + \alpha\upsilon\omega \qquad (4.2)$$

but now with three unknown parameters multiplied together. The unknown scales are υ and ω, while α is a regression parameter estimated once the scales are calculated.

Here we shall take a much simpler table as illustration, the relationship between length of stay for schizophrenic patients in London mental hospitals and frequency of visit (Table 4.2).

Years	Goes Home or Visited Regularly	Visited Less Than Once a Month and Does Not Go Home	Never Visited and Never Goes Home
2-10	43	6	9
10-20	16	11	18
>20	3	10	16

Table 4.2 Schizophrenic Patients in London (Fienberg, 1977, p.55)

The macro L2OV fits a series of models, the first of which is again independence:

```
SCHIZOPHRENIC PATIENTS IN LONDON (FIENBERG, 1977, P.55)

 Independence Model

 scaled deviance = 38.353 at cycle  4
            d.f. =  4

         estimate          s.e.      parameter
     1      3.305        0.1606      1
     2    -0.8313        0.2305      VIS(2)
     3    -0.3659        0.1984      VIS(3)
     4    -0.2538        0.1987      LENG(2)
     5    -0.6931        0.2273      LENG(3)
     scale parameter taken as  1.000
```

This model must be rejected. Frequency of visit depends on years of internment; the question is, how?

The second model takes both variables as metric and linear, i.e. as equally spaced scales.

```
SCHIZOPHRENIC PATIENTS IN LONDON (FIENBERG, 1977, P.55)

 Linear Effects Model

 scaled deviance = 7.1192 at cycle  3
            d.f. = 3

         estimate          s.e.      parameter
     1      2.894        0.1820      1
     2    -0.5215        0.2506      VIS(2)
     3   -0.08883        0.2301      VIS(3)
     4  -0.007450        0.2193      LENG(2)
     5    -0.6458        0.2574      LENG(3)
     6     0.1971       0.03962      ZZ6_
     scale parameter taken as  1.000
```

Again, the fit is improved, but just barely acceptable. The macro has created the new variable, ZZ6_, which is the product of the two linear variables. The parameter estimate is the slope, which, being positive, shows that the two variables vary together. However, frequency of visit *decreases* from left to right in the table, so that it decreases with increasing length of stay.

The following two models fit, the first successfully, one variable as metric and linear and the other as nominal, then the reverse:

```
SCHIZOPHRENIC PATIENTS IN LONDON (FIENBERG, 1977, P.55)

 Column (K) Effect Model

 scaled deviance = 0.019817 at cycle  3
           d.f. = 2

         estimate          s.e.       parameter
    1       2.129         0.3173      1
    2     -0.3743         0.2769      VIS(2)
    3     0.08700         0.2500      VIS(3)
    4      0.6561         0.3097      LENG(2)
    5      0.5674         0.4096      LENG(3)
    6     -0.8152         0.1653      VIS(1).ZZ4_
    7    -0.01298         0.1634      VIS(2).ZZ4_
    8       0.000        aliased      VIS(3).ZZ4_
    scale parameter taken as  1.000

SCHIZOPHRENIC PATIENTS IN LONDON (FIENBERG, 1977, P.55)

 Row (I) Effect Model

 scaled deviance = 6.4589 at cycle  3
           d.f. = 2

         estimate          s.e.       parameter
    1       2.175         0.3134      1
    2      0.1653         0.3541      VIS(2)
    3       1.276         0.4857      VIS(3)
    4     0.02758         0.2283      LENG(2)
    5     -0.5594         0.2749      LENG(3)
    6     -0.7668         0.1589      LENG(1).YY4_
    7     -0.2899         0.1483      LENG(2).YY4_
    8       0.000        aliased      LENG(3).YY4_
    scale parameter taken as  1.000
```

In the first of these models, when length of stay is metric and linear, the fit is very acceptable, while it is not in the second, where frequency of visit has the equal interval scale. YY4_ and ZZ4_ are the new linear variables created by the macro. In the column effect model, we see that longer length of stay is less probable for the first category of visits (-0.815) as compared to the other two categories with more or less zero slopes.

A first row and column effect model combines the previous two models and, of course, fits well since the first of these did.

```
SCHIZOPHRENIC PATIENTS IN LONDON (FIENBERG, 1977, P.55)

 Row and Column Effect Model (1)

 scaled deviance = 0.0024777 at cycle  3
           d.f. = 1

         estimate          s.e.       parameter
    1       2.148         0.3467      1
    2     -0.3618         0.2927      VIS(2)
    3      0.1182         0.3441      VIS(3)
```

```
     4      0.6217        0.4051      LENG(2)
     5      0.5085        0.6069      LENG(3)
     6     -0.7743        0.3495      VIS(1).ZZ4_
     7    0.001469        0.1969      VIS(2).ZZ4_
     8       0.000      aliased       VIS(3).ZZ4_
     9    -0.03278        0.2485      LENG(1).YY4_
    10       0.000      aliased       LENG(2).YY4_
    11       0.000      aliased       LENG(3).YY4_
     scale parameter taken as   1.000
```

However, it is not acceptable, since it is more complex than that with length of stay on a linear metric scale.

Finally, the model with the two ordinal variable scales estimated is fitted.

```
SCHIZOPHRENIC PATIENTS IN LONDON (FIENBERG, 1977, P.55)

 Row and Column Effect Model (2)

 Scale for First Ordinal Variable
   0.       0.9837    1.000

 Scale for Second Ordinal Variable
   0.       0.5116    1.000

 scaled deviance =    0.0115 at cycle  0.
            d.f. =    1.

          estimate           s.e.        parameter
     1       3.761         0.1505        1
     2      -1.986         0.3190        VIS(2)
     3      -1.552         0.3019        VIS(3)
     4     -0.9871         0.2459        LENG(2)
     5      -2.666         0.4895        LENG(3)
     6       3.236         0.6062        ZZ6_
     scale parameter taken as   1.000

 SCHIZOPHRENIC PATIENTS IN LONDON (FIENBERG, 1977, P.55)

 Analysis of Association Table

     Effect                  Chi2    df    Prob
 General Effect             31.23    1.   0.
 Column Effect              7.099    1.   0.0077
 Row Effect                0.6603    1.   0.4165
 Other Effects(1)          0.0025    1.   0.9603
 Other Effects(2)          0.0115    1.   0.9145
```

The analysis shows that, as already concluded, the scale for length of stay has virtually equal intervals, while we now see clearly what we suspected from the column effect model: the scale for frequency of visit contrasts "goes home or visited regularly" with the two cases of few or no visits and not going home. These results are summarized in the analysis of association table which partitions the Chi-squares. The "general effect" refers to the lack of independence between the two variables while the "column effect" refers to the lack of linearity of the frequency of visit variable. Both are significant, as we have seen. "Row effect" tests for linearity of length of stay. The two "other effects" are interchangeable and only one should be interpreted. They concern any lack of fit remaining when both row and column effects are included in the model.

We shall now study the same table with two other approaches to ordinal variables.

3. The Proportional Odds Model

Since the categories of an ordinal variable are, by definition, ordered, frequencies of response in successive categories may be compared. This and the following section present two common approaches to such comparisons.

In the proportional odds model, each category is considered in turn and the frequency of response at least up to that point on the ordinal scale is compared to the frequency for all points higher on the scale. The first category is compared to all the rest combined, then the first and second combined are compared to all the rest combined, and so on. In this way, the original table with a K category ordinal scale is converted into a series of K-1 subtables, each with a binary categorization, lower/higher than the point on the scale. We then have three types of variables, the binary dependent variable indicating more or less on the ordinal scale, a variable indexing the subtables corresponding to the points on the ordinal scale, and the explanatory variables.

It might appear from the construction of this table that we now have a simple case where the logistic model could be applied to the binary response variable. However, if the observations in the original table were independent, the categories in the new reconstructed table no longer will be. A more complex analysis is called for, one which does not fall into the standard GLIM distributions. A series of macros are necessary to define the resulting distribution, with what is known as a composite link function. GLIM provides facilities for such applications, but the technical details are beyond the scope of this book. The analysis is set up and applied through the macro POOV. One restriction is that the explanatory variables may not be defined by $FActor. If such a setup is required, the macro TRAN may be used instead.

We now apply the proportional odds model to our data on schizophrenic patients of the previous section. The macro first prints out the reconstructed table, with ZZ1_ the index of subtables and ZZ2_ the rows of the (complex) explanatory variable, then the approximate analysis using the binomial distribution and the logistic model, and finally the proportional odds model. Since we have already seen that length of stay may be treated as an equal interval (linear metric) scale, we use that here.

SCHIZOPHRENIC PATIENTS IN LONDON (FIENBERG, 1977, P.55)

```
 Proportional Odds Model

       R           N        ZZ1_      ZZ2_
    43.000      58.00     1.000     1.000
    16.000      45.00     1.000     2.000
     3.000      29.00     1.000     3.000
    49.000      58.00     2.000     1.000
    27.000      45.00     2.000     2.000
    13.000      29.00     2.000     3.000

 Approximate Analysis

 scaled deviance = 3.5486 at cycle  3
            d.f. = 3

 Chi2 probability =   0.3143 for Chi2 =    3.549 with    3. d.f.

          estimate          s.e.     parameter
     1     -0.4543         0.2029     C1_
     2      0.5948         0.2058     C2_
     3     -3.089          0.4745     C3_
     4      0.000         aliased     C4_
     5      0.000         aliased     C5_
     6      0.000         aliased     C6_
     7      0.000         aliased     C7_
     8      0.000         aliased     C8_
     9      0.000         aliased     C9_
     scale parameter taken as   1.000

 Exact Analysis

 scaled deviance = 6.6864 at cycle  2
            d.f. = 6

 Chi2 probability =   0.3507 for Chi2 =    6.686 with    6. d.f.

          estimate          s.e.     parameter
     1     -0.3863         0.1994     C1_
     2      0.6635         0.2051     C2_
     3     -3.047          0.5859     C3_
     4      0.000         aliased     C4_
     5      0.000         aliased     C5_
     6      0.000         aliased     C6_
     7      0.000         aliased     C7_
     8      0.000         aliased     C8_
     9      0.000         aliased     C9_
     scale parameter taken as   1.000
```

The model fits very well. The negative value for C3_ indicates that the odds of receiving more rather than less visits decreases with increasing length of stay, confirming the previous results. C1_ and C2_ give the parameters for the two subtables, which are not interpreted.

4. The Continuation Ratio Model

The continuation ratio model resembles closely the proportional odds model. A series of subtables is also constructed here. But now for each category of the ordinal variable considered in turn, the frequency of response at least up to that point on the ordinal scale is compared only to the frequency for the immediately following category. The first category is compared to the second, the first and second combined to the third, and so on. Given that the response is at least at a given level, what is the chance of continuing to the immediately following level? Again, the original table with a K category ordinal scale is converted into a series of K-1 subtables. However, here the macro, CROV, requires that all explanatory variables be treated as one complex variable, in the same way as in the first section of this chapter for the macro L1OV. Then again, as for the proportional odds model, we have three distinct types of variables: the binary dependent variable, the variable indexing subtables, and the (complex) explanatory variable.

In contrast to the proportional odds model, with the continuation ratio model, independence is retained when the table is reconstructed, and the logistic model may be directly applied. The macro, CROV, simply reconstructs the table and fits this model. When applied to our schizophrenic data, the model fits very well:

```
SCHIZOPHRENIC PATIENTS IN LONDON (FIENBERG, 1977, P.55)

 Continuation Ratio Model

        R          N        ZZ1_       ZZ2_
     43.000     49.00     1.000      1.000
     16.000     27.00     1.000      2.000
      3.000     13.00     1.000      3.000
     49.000     58.00     2.000      1.000
     27.000     45.00     2.000      2.000
     13.000     29.00     2.000      3.000

 scaled deviance = 1.9695 at cycle   4
             d.f. = 2

 Chi2 probability =   0.3735 for Chi2 =    1.969 with    2. d.f.

          estimate          s.e.        parameter
     1        1.745        0.3260       1
     2       0.1287        0.3264       ZZ1_(2)
     3       -1.432        0.3694       ZZ2_(2)
     4       -2.320        0.4266       ZZ2_(3)
     scale parameter taken as    1.000
```

However, here we have length of stay as a nominal factor variable. In order to compare our results with those for the proportional odds model of the previous section, we refit the continuation ratio table with the equal interval scale for length of stay. We keep the variable ZZ1_, which indexes the tables, and replace the new complex variable, ZZ2_, created by the macro, with the linear orthogonal polynomial.

```
SCHIZOPHRENIC PATIENTS IN LONDON (FIENBERG, 1977, P.55)

 scaled deviance = 2.6856 at cycle  4
            d.f. = 3

 Chi2 probability =  0.4447 for Chi2 =   2.686 with   3. d.f.

          estimate         s.e.      parameter
     1      0.4700       0.2557      1
     2      0.1261       0.3261      ZZ1_(2)
     3     -2.357        0.4220      LENL
     scale parameter taken as   1.000
```

The result is very similar to that for the previous section, with the same interpretation.

The choice among the three approaches to the analysis of ordinal variables presented in this chapter is rarely obvious. As seen here, the three are often mutually reinforcing, and not all would be necessary in most situations. The log-multiplicative model is often attractive because it provides a scale. The continuation ratio model is perhaps most specialized, being applicable where one is interested in continuation to each successively higher point on a scale, but it is the most easily fitted in terms of computing time.

CHAPTER 5

ZERO FREQUENCIES AND INCOMPLETE TABLES

1. Sampling Zeroes

When several categorical variables are cross-tabulated to form a table of several dimensions, some cells often contain zero frequencies of response. Such a situation may arise in at least three ways. The zero may have occurred simply because the sample is not large enough and that combination of categories is not represented. A large enough sample would theoretically include such combinations. In the other two cases, the combination of categories is actually impossible or is excluded from the model for some theoretical reason. In the first case, the expected frequencies or fitted values should be positive. In the latter two cases, the expected frequencies for any model must be zero. In this section, we consider the first case, that of sampling zeroes. In subsequent sections, various possibilities of incomplete tables with structural zeroes will be covered.

If a saturated model is to be fitted, even one sampling zero will create problems for estimation of the parameters. GLIM will only estimate as many parameters as there are non-zero entries in the table. The last parameters encountered in the list will not be estimated, even although they may have no actual relationship to the location of the zeroes in the table.

For an unsaturated model, all of the parameters may most often be estimated, as long as the number of non-zero frequencies is at least as large as the number of parameters to be estimated. However, unexpected exceptions to this may occur, depending on the location of the zeroes in the table. GLIM provides no systematic warning of a problem being encountered, but goes ahead and estimates the parameters anyway. The user must beware, especially if GLIM has also tabulated the table so that it has not been inspected beforehand. The most reliable indication that a problem has occurred is if certain standard errors of parameter estimates (with $Display E) are greatly inflated. This does not, however, indicate which zeroes are causing the problem. The solution is to eliminate those zero cells for which the estimated frequencies (given by $Display R) are either very large or very small. (Remember that, in the situation of sampling zeroes, the estimated frequencies should be positive.) These cells are removed by giving them a weight zero. In this way, the degrees of freedom are reduced (corrected) by the number of cells eliminated. Note that the cells to be eliminated will depend on the model fitted to the data and not simply on the location of the zeroes in the table. A macro, DFCT, performs the two tests and the elimination automatically.

A simple example illustrates the problem. Table 5.1 gives changes in vote between the Swedish elections of 1964 and 1970. No one changed between Communist and Conservative, in either direction, so that the table contains two sampling zeroes.

	Communist	Social Democrat	Centre	People's Party	Conservative
Communist	22	27	4	1	0
Social Democrat	16	861	57	30	8
Centre	4	26	248	14	7
People's Party	8	20	61	201	11
Conservative	0	4	31	32	140

Table 5.1 Swedish Elections, 1964 and 1970 (Fingleton, 1984, p.138)

We first apply a saturated model, which we know GLIM cannot fit completely.

```
SWEDISH ELECTIONS 1964 AND 1970 (FINGLETON, 1984, P.138)

scaled deviance =******** at cycle  9
          d.f. = 1
    (change in d.f.)

          estimate          s.e.      parameter
    1        3.091        0.2132      1
    2       0.2049        0.2872      V70(2)
    3       -1.705        0.5436      V70(3)
    4       -3.091         1.022      V70(4)
    5       0.9253         54.60      V70(5)
    6      -0.3183        0.3286      V64(2)
    7       -1.705        0.5436      V64(3)
    8       -1.011        0.4129      V64(4)
    9       0.9253         54.60      V64(5)
   10        3.781        0.3823      V70(2).V64(2)
   11        1.667        0.6091      V70(2).V64(3)
   12       0.7114        0.5074      V70(2).V64(4)
   13       -2.835         54.60      V70(2).V64(5)
   14        2.975        0.6128      V70(3).V64(2)
   15        5.832        0.7413      V70(3).V64(3)
   16        3.736        0.6609      V70(3).V64(4)
   17        1.122         54.60      V70(3).V64(5)
   18        3.720         1.068      V70(4).V64(2)
   19        4.344         1.169      V70(4).V64(3)
   20        6.315         1.084      V70(4).V64(4)
   21        2.540         54.61      V70(4).V64(5)
   22       -1.618         54.60      V70(5).V64(2)
   23      -0.3657         54.60      V70(5).V64(3)
   24      -0.6069         54.60      V70(5).V64(4)
   25        0.000       aliased      V70(5).V64(5)
    scale parameter taken as  1.000

SWEDISH ELECTIONS 1964 AND 1970 (FINGLETON, 1984, P.138)

  unit  observed    fitted   residual
     1        22    21.998      0.000
     2        27    27.000      0.000
     3         4     4.000      0.000
     4         1     1.000      0.000
     5         0    55.494     -7.449
     6        16    16.000      0.000
     7       861   861.000      0.000
```

```
  8        57     57.000       0.000
  9        30     30.000       0.000
 10         8      8.000       0.000
 11         4      4.000       0.000
 12        26     26.000       0.000
 13       248    248.000       0.000
 14        14     14.000       0.000
 15         7      7.000       0.000
 16         8      8.000       0.000
 17        20     20.000       0.000
 18        61     61.000       0.000
 19       201    201.000       0.000
 20        11     11.000       0.000
 21         0     55.494      -7.449
 22         4      4.000       0.000
 23        31     31.000       0.000
 24        32     32.000       0.000
 25       140    139.998       0.000
```

We see that 8 parameters have inflated standard estimates while only one is not estimated. An additional indication of the problem, one which occurs less frequently, is the loss of degrees of freedom: one degree of freedom remains when none should. The two zero cells both have very large fitted values. We apply the macro to eliminate the two cells.

```
SWEDISH ELECTIONS 1964 AND 1970 (FINGLETON, 1984, P.138)

Model with corrected df

-- model changed

scaled deviance = 4.983e-11 at cycle  1
            d.f. = 0          from 23 observations

          estimate          s.e.       parameter
  1          3.091        0.2132       1
  2         0.2048        0.2872       V70(2)
  3         -1.705        0.5436       V70(3)
  4         -3.091         1.022       V70(4)
  5         0.3185        0.4647       V70(5)
  6        -0.3185        0.3286       V64(2)
  7         -1.705        0.5436       V64(3)
  8         -1.012        0.4129       V64(4)
  9          1.532        0.5182       V64(5)
 10          3.781        0.3823       V70(2).V64(2)
 11          1.667        0.6091       V70(2).V64(3)
 12         0.7115        0.5074       V70(2).V64(4)
 13         -3.442        0.7454       V70(2).V64(5)
 14          2.975        0.6128       V70(3).V64(2)
 15          5.832        0.7413       V70(3).V64(3)
 16          3.736        0.6609       V70(3).V64(4)
 17         0.5155        0.7421       V70(3).V64(5)
 18          3.720         1.068       V70(4).V64(2)
 19          4.344         1.169       V70(4).V64(3)
 20          6.315         1.084       V70(4).V64(4)
 21          1.934         1.140       V70(4).V64(5)
 22         -1.012        0.6351       V70(5).V64(2)
 23         0.2412        0.7802       V70(5).V64(3)
```

```
   24        0.000      aliased      V70(5).V64(4)
   25        0.000      aliased      V70(5).V64(5)
    scale parameter taken as  1.000
```

SWEDISH ELECTIONS 1964 AND 1970 (FINGLETON, 1984, P.138)

unit	observed	fitted	residual
1	22	22.000	0.000
2	27	27.000	0.000
3	4	4.000	0.000
4	1	1.000	0.000
5	0	30.250	0.000
6	16	16.000	0.000
7	861	861.000	0.000
8	57	57.000	0.000
9	30	30.000	0.000
10	8	8.000	0.000
11	4	4.000	0.000
12	26	26.000	0.000
13	248	248.000	0.000
14	14	14.000	0.000
15	7	7.000	0.000
16	8	8.000	0.000
17	20	20.000	0.000
18	61	61.000	0.000
19	201	201.000	0.000
20	11	11.000	0.000
21	0	101.818	0.000
22	4	4.000	0.000
23	31	31.000	0.000
24	32	32.000	0.000
25	140	140.000	0.000

GLIM does not estimate the last two parameters. Note that those estimates which already had small standard errors have changed very little after the correction.

Consider now a model where the parties are placed on an equal interval scale. This model is introduced primarily for illustrative purposes and is not meant to be realistic. If we actually wished to place the parties on an ordered scale, we should consider the log-multiplicative model. We fit a model with linear interaction between the votes at the two elections.

SWEDISH ELECTIONS 1964 AND 1970 (FINGLETON, 1984, P.138)

```
scaled deviance = 665.42 at cycle  5
           d.f. =  15
```

	estimate	s.e.	parameter
1	0.7553	0.1820	1
2	-0.1618	0.1892	V70(2)
3	-4.812	0.3161	V70(3)
4	-10.64	0.5521	V70(4)
5	-18.08	0.8596	V70(5)
6	-0.2462	0.1864	V64(2)
7	-5.206	0.3180	V64(3)
8	-10.26	0.5305	V64(4)
9	-17.16	0.8195	V64(5)

```
10        1.570      0.06681      V704
  scale parameter taken as  1.000

SWEDISH ELECTIONS 1964 AND 1970 (FINGLETON, 1984, P.138)

   unit  observed     fitted   residual
      1        22     10.229      3.680
      2        27     41.823     -2.292
      3         4      1.921      1.500
      4         1      0.027      5.886
      5         0      0.000     -0.009
      6        16     38.435     -3.619
      7       861    755.285      3.847
      8        57    166.733     -8.498
      9        30     11.393      5.513
     10         8      0.154     20.004
     11         4      1.296      2.375
     12        26    122.406     -8.714
     13       248    129.876     10.365
     14        14     42.654     -4.387
     15         7      2.768      2.543
     16         8      0.040     39.905
     17        20     18.064      0.455
     18        61     92.121     -3.242
     19       201    145.414      4.610
     20        11     45.361     -5.102
     21         0      0.000     -0.014
     22         4      0.422      5.506
     23        31     10.349      6.420
     24        32     78.513     -5.249
     25       140    117.716      2.054
```

This time, the standard errors are not inflated and it is not clear whether the two sampling zeroes are affecting the model fit. However, the estimated frequencies are very small for the two zero cells, so we eliminate them by applying the macro.

```
SWEDISH ELECTIONS 1964 AND 1970 (FINGLETON, 1984, P.138)

Model with corrected df

-- model changed
scaled deviance = 665.42 at cycle  2
           d.f. = 13     from 23 observations

          estimate         s.e.       parameter
   1        0.7553       0.1820       1
   2       -0.1617       0.1893       V70(2)
   3        -4.812       0.3164       V70(3)
   4        -10.64       0.5528       V70(4)
   5        -18.08       0.8606       V70(5)
   6       -0.2462       0.1865       V64(2)
   7        -5.206       0.3183       V64(3)
   8        -10.26       0.5311       V64(4)
   9        -17.16       0.8205       V64(5)
  10         1.570      0.06689       V704
  scale parameter taken as  1.000
```

SWEDISH ELECTIONS 1964 AND 1970 (FINGLETON, 1984, P.138)

unit	observed	fitted	residual
1	22	10.229	3.680
2	27	41.823	-2.292
3	4	1.921	1.500
4	1	0.027	5.886
5	0	0.000	0.000
6	16	38.435	-3.619
7	861	755.285	3.847
8	57	166.733	-8.498
9	30	11.393	5.513
10	8	0.154	20.004
11	4	1.296	2.375
12	26	122.406	-8.714
13	248	129.876	10.365
14	14	42.654	-4.387
15	7	2.768	2.543
16	8	0.040	39.904
17	20	18.064	0.455
18	61	92.121	-3.242
19	201	145.413	4.610
20	11	45.361	-5.102
21	0	0.000	0.000
22	4	0.422	5.506
23	31	10.349	6.420
24	32	78.513	-5.249
25	140	117.716	2.054

The deviance remains unchanged, but the degrees of freedom are reduced by two.

It cannot be emphasized too much that great care must be taken with sampling zeroes when using GLIM. Several of the examples in this and the following chapter contain them. They have been left uncorrected for the reader to discover.

2. Incomplete Tables and Quasi-Independence

When a table involves structural zeroes, these are simply not included in the data and GLIM fits the model without problem. Note however that %GL may not be used to calculate the variables, since the table is no longer symmetric.

Since, with variables defined by $FActor, GLIM performs the analysis automatically, it will be more useful to apply the macro TRAN; as well, this will provide more easily interpretable estimates. Our example (Table 5.2) involves health problems, sex, and age of young people. The combination male with menstruation problems is impossible.

The base model has the three sets of mean parameters and the sex-age interaction between the two independent variables. When some cells are missing in a table, such a model is called quasi-independence.

Sex	Age	Health Problem Sex & Reproduction	Menstruation	How Healthy I Am	Nothing
Male	12-15	4	-	42	57
Male	16-17	2	-	7	20
Female	12-15	9	4	19	71
Female	16-17	7	8	10	31

Table 5.2 Health Problems of Young People (Fienberg, 1977, p.116)

```
HEALTH PROBLEMS (FIENBERG, 1977, P.116)

scaled deviance = 22.025 at cycle  4
            d.f. =  7

Chi2 probability =  0.0026 for Chi2 =   22.02 with   7. d.f.

        estimate          s.e.       parameter
  1       2.435         0.1026       1
  2     -0.8492         0.1792       PRO1
  3     -0.8145         0.2295       PRO2
  4      0.4165         0.1257       PRO3
  5     -0.1253        0.06792       SEX
  6      0.4692        0.06697       AGE
  7      0.1645        0.06697       SA

  scale parameter taken as  1.000
```

The model is rejected; type of health problem depends on age or sex or both.

We introduce the effect of sex. Note that sex cannot interact with the second type of problem, menstruation, so this is not included in the model.

```
HEALTH PROBLEMS (FIENBERG, 1977, P.116)

scaled deviance = 9.4260 (change =   -12.60) at cycle  4
            d.f. = 5      (change =    -2   )

Chi2 probability =  0.0922 for Chi2 =   9.426 with   5. d.f.

Chi2 probability =  0.0018 for Chi2 =   12.60 with   2. d.f.

        estimate          s.e.       parameter
  1       2.379         0.1119       1
  2     -0.9074         0.1892       PRO1
  3     -0.8275         0.2314       PRO2
  4      0.4400         0.1377       PRO3
  5     -0.1944        0.09768       SEX
  6      0.4692        0.06697       AGE
  7      0.1645        0.06697       SA
  8     -0.3675         0.1662       SP1
  9      0.3852         0.1144       SP3
  scale parameter taken as  1.000
```

The model now fits satisfactorily and is a significant improvement on complete quasi-independence. Boys have fewer problems of sex and reproduction (-0.3675) than girls, and relatively more with general health (0.3852), while about the same proportion of each sex state that they have no problem (0.3675 - 0.3852 = -0.0177).

We continue by replacing the effect of sex by that of age.

```
HEALTH PROBLEMS (FIENBERG, 1977, P.116)

 scaled deviance = 13.447 (change =  +4.021) at cycle  3
          d.f. =  4       (change =  -1     )

 Chi2 probability =  0.0094 for Chi2 =   13.45 with   4. d.f.

 Chi2 probability =  0.0349 for Chi2 =   8.577 with   3. d.f.

         estimate          s.e.       parameter
    1       2.434        0.1054       1
    2     -0.7616        0.1846       PRO1
    3     -0.8170        0.2431       PRO2
    4      0.3293        0.1414       PRO3
    5     -0.1161       0.06851       SEX
    6      0.2797        0.1054       AGE
    7      0.1359       0.06851       SA
    8    -0.08026        0.1846       AP1
    9     -0.4904        0.2431       AP2
   10      0.3747        0.1414       AP3
      scale parameter taken as   1.000
```

This model does not fit sufficiently well and must be rejected. Apparently, the type of health problem does not depend on age, at least for the two age groups considered here. It is not necessary to continue. Our second model, with sex, but not age, influencing health problems, will be retained. Males have relatively fewer problems with sex and reproduction and more with general health.

Note that if the interaction between sex and age with respect to health problems were to be included, it would also contain two parameters, as for sex with problems, since, again, the second problem does not interact with sex.

3. Population Estimation

A special case of an incomplete table occurs with the problem of population estimation. Suppose that we have several estimators of a population and that we know which individuals are touched by each estimator. Then we have a series of dichotomous variables indicating whether or not an individual is included in each estimator. The cell categorizing all individuals not included in any estimator is missing. We wish to estimate the frequency in this missing cell and, hence, obtain an estimate of the total population. Since we cannot estimate a saturated model when one category is missing, we must make some assumption of independence among the different types of estimators used. Under high order independence

$$\frac{n_{111}n_{122}n_{212}n_{221}}{n_{222}n_{112}n_{121}n_{211}} = 1 \qquad (5.1)$$

for example, with three population estimates. Then, if n_{222} is the frequency for the missing cell,

$$n_{222} = \frac{n_{111}n_{122}n_{212}n_{221}}{n_{122}n_{121}n_{211}} \tag{5.2}$$

Since we shall require an interval of plausible values for the total population estimate, N, we need an estimate of the variance (or standard deviation):

$$\text{var}(N) = \frac{N\, n_{222}}{n_{112}+n_{121}+n_{211}+n_{111}} \tag{5.3}$$

These quantities are easily obtained with GLIM. The missing frequency may be obtained directly from the fitted values if any arbitrary value is supplied and given a zero weight for the fit. The variance must be calculated from the formula (5.3).

As an example, we take the estimation of the number of formal volunteer organizations in Massachusetts towns (Table 5.3). The three estimators are newspapers, telephones, and a census.

Newspaper	Telephone	Census Yes	Census No
Yes	Yes	4	1
No	Yes	8	2
Yes	No	16	49
No	No	113	-

Table 5.3 Estimation of the Number of Formal Volunteer Organizations (Bishop et al, 1975, p.243)

The model with no three-factor interaction fits perfectly and we easily discover that the interaction between census and newspapers may be eliminated. We fit this model and use a specially constructed macro to calculate the population estimation and its standard deviation.

```
FORMAL VOLUNTEER ORGANIZATIONS (BISHOP ET AL, 1975, P.243)

scaled deviance = 0.0000000 at cycle  5
          d.f. = 1

Chi2 probability =   1.000 for Chi2 =  0.     with   1. d.f.

       estimate          s.e.        parameter
  1       1.386        0.4655        1
  2      -1.386        0.6455        CENS(2)
  3      0.6931        0.5477        NEWS(2)
  4       1.386        0.5284        TELE(2)
  5       2.506        0.7068        CENS(2).TELE(2)
  6       1.262        0.6094        NEWS(2).TELE(2)
  scale parameter taken as   1.000

 unit  observed     fitted    residual
   1          4      4.000      0.0000
```

```
      2          1      1.000      0.0000
      3          8      8.000      0.0000
      4          2      2.000      0.0000
      5         16     16.000      0.0000
      6         49     49.000      0.0000
      7        113    113.000      0.0000
      8          0    346.100      0.0000

Estimated total =     539. with s.d. =     80.
```

From the fitted values, we see that the number of missing organisations is estimated as 346. With asymptotic normality, a 95% interval covers two standard deviations, giving (379, 699) for the total number of organisations. If we also eliminate the newspaper/telephone interaction, the model is still very acceptable:

```
scaled deviance = 3.8244 (change =   +3.824) at cycle  4
          d.f. = 2       (change =   +1     )

Chi2 probability =  0.1478 for Chi2 =    3.824 with    2. d.f.

        estimate          s.e.       parameter
    1     0.5596        0.3521       1
    2     -1.386        0.6455       CENS(2)
    3      1.768        0.2361       NEWS(2)
    4      2.375        0.3018       TELE(2)
    5      2.344        0.6967       CENS(2).TELE(2)
    scale parameter taken as   1.000

  unit  observed     fitted    residual
     1         4      1.750       1.701
     2         1      0.437       0.850
     3         8     10.250      -0.702
     4         2      2.562      -0.351
     5        16     18.810      -0.648
     6        49     49.000       0.000
     7       113    110.200       0.268
     8         0    287.000       0.000

Estimated total =     480. with s.d. =    66.
```

One important result of simplifying the model is that the standard deviation is always smaller (Bishop et al, 1975, p.242). Of course, we must still keep an acceptable model. Here our interval becomes (347, 611). Further simplification of the model, by eliminating the census/telephone interaction, is not possible.

The parameter estimates indicate that there is a positive association between being covered by the census and being listed in the telephone directory, while coverage by newspapers is relatively independent of both of these.

4. Social Mobility

A series of standard social mobility models have been described by Duncan (1979). These may all easily be fitted with GLIM. The most important ones involve elimination of specific cells from the table for theoretical reasons, yielding an incomplete table.

A mobility table is a square two-dimensional table with the same categorical variable observed at two points in time. If the same individuals are involved, it is a form of panel study. Examples include the British social mobility table of Chapter 2 and the voting change table of the first section of this chapter; the second is a panel study, as is a further common example, used here, migrant behaviour (Table 5.4).

	1971			
1966	Central Clydesdale	Urban Lancashire and Yorkshire	West Midlands	Greater London
Central Clydesdale	118	12	7	23
Urban Lanc. & York.	14	2127	86	130
West Midlands	8	69	2548	107
Greater London	12	110	88	7712

Table 5.4 Migrant Behaviour in Britain between 1966 and 1971 (Fingleton, 1984, p.142)

We wish to test if position at the second point in time depends on that at the first point in time, whether it be profession, vote, place of residence, or whatever. This is the standard model of independence which we have encountered many times. However, here, as we have already noted with such tables, the problem is that too many individuals do not change position between the two time points for such independence to be acceptable. Too many observations appear on the diagonal. The simple solution is to eliminate these diagonal elements and test for quasi-independence. More theoretically, this approach assumes that the diagonal contains two types of individuals, the movers, who might have moved, but did not happen to in the observed time interval, and the stayers who never change. Hence, the name of the model: the mover-stayer model.

Duncan's (1979) other standard mobility models assume an ordering for the categories and fit an equal interval scale. Although the five models may easily be fitted by the usual GLIM procedures, for facility, they have been assembled in a single macro called SMCT. We apply the macro to the migration table, where we note that the geographical locations are ordered from north to south of Britain. We also observe the exceptionally high values on the diagonal.

The first model fitted is the usual one for independence:

```
MIGRANT BEHAVIOUR - FINGLETON (1984, P.142)

1. Independence Model

scaled deviance = 19884. at cycle  6
          d.f. =     9

Chi2 probability =  0.     for Chi2 =   19884. with   9. d.f.
```

```
        estimate          s.e.       parameter
  1       0.6134        0.1126       1
  2        2.690       0.08165       M66(2)
  3        2.838       0.08129       M66(3)
  4        3.902       0.07981       M66(4)
  5        2.725       0.08327       M71(2)
  6        2.888       0.08288       M71(3)
  7        3.960       0.08141       M71(4)
  scale parameter taken as   1.000

unit  observed     fitted    residual
   1       118      1.847      85.476
   2        12     28.159      -3.045
   3         7     33.152      -4.542
   4        23     96.843      -7.504
   5        14     27.203      -2.531
   6      2127    414.815      84.067
   7        86    488.365     -18.207
   8       130   1426.619     -34.329
   9         8     31.531      -4.191
  10        69    480.812     -18.781
  11      2548    566.064      83.302
  12       107   1653.595     -38.033
  13        12     91.430      -8.307
  14       110   1394.214     -34.393
  15        88   1641.420     -38.342
  16      7712   4794.941      42.126
```

As would be expected, this model is definitely not acceptable. We note the very large underestimation of all diagonal cells.

The second model, called row effects (assuming that the first time point forms the rows of the table) takes the second position as a linear equal interval scale and the first position as a nominal variable and fits the interaction between them.

```
MIGRANT BEHAVIOUR - FINGLETON (1984, P.142)

2. Row Effects Model

scaled deviance = 4155.6 (change =   -15728.) at cycle  6
           d.f. =      6  (change =        -3 )

Chi2 probability =  0.      for Chi2 =    4156. with    6. d.f.

        estimate          s.e.       parameter
  1        4.194        0.1217       1
  2       0.2487        0.1603       M66(2)
  3       -4.568        0.2012       M66(3)
  4       -15.75        0.2892       M66(4)
  5        11.00        0.1588       M71(2)
  6        17.40        0.2346       M71(3)
  7        20.49        0.2647       M71(4)
  8       -10.66        0.2191       M66(1).ZZ1_
  9       -7.911        0.1125       M66(2).ZZ1_
 10       -4.717       0.07763       M66(3).ZZ1_
 11        0.000      aliased        M66(4).ZZ1_
  scale parameter taken as   1.000
```

```
unit  observed     fitted   residual
   1       118     66.302      6.349
   2        12     92.400     -8.364
   3         7      1.301      4.996
   4        23      0.001    887.022
   5        14     85.026     -7.703
   6      2127   1858.305      6.233
   7        86    410.360    -16.012
   8       130      3.325     69.466
   9         8      0.688      8.814
  10        69    366.729    -15.547
  11      2548   1974.478     12.907
  12       107    390.105    -14.334
  13        12      0.000   3883.525
  14       110      0.569    145.045
  15        88    342.862    -13.764
  16      7712   7578.569      1.533
```

Although a slight improvement, the model is still not acceptable. The diagonal estimates are, however, much better. The parameter estimates are the slopes for each category of origin, each calculated in relation to the last category. As can also be seen from the table, all slopes are negative relative to the last line of the table. However, migration both ways between Clydesdale and London is especially underestimated. This is due to the linear scale which should continue to decrease from Clydesdale, through Lancashire, Yorkshire, and the West Midlands to London, but increases for London in the table.

The macro now gives a zero weight to the diagonal elements and refits the (quasi-) independence model:

```
MIGRANT BEHAVIOUR - FINGLETON (1984, P.142)

3. Quasi-independence (Mover-Stayer) Model

scaled deviance = 4.3666 at cycle  3
           d.f. = 5        from 12 observations

Chi2 probability =  0.4994 for Chi2 =    4.367 with   5. d.f.

        estimate          s.e.      parameter
   1      0.4615        0.2345      1
   2       2.010        0.1705      M66(2)
   3       1.724        0.1728      M66(3)
   4       2.124        0.1742      M66(4)
   5       2.085        0.1890      M71(2)
   6       1.914        0.1886      M71(3)
   7       2.455        0.1865      M71(4)
   scale parameter taken as  1.000

unit  observed     fitted   residual
   1       118      1.586      0.000
   2        12     12.758     -0.212
   3         7     10.757     -1.145
   4        23     18.485      1.050
   5        14     11.836      0.629
   6      2127     95.185      0.000
   7        86     80.252      0.642
```

```
  8        130    137.912      -0.674
  9          8      8.892      -0.299
 10         69     71.505      -0.296
 11       2548     60.287       0.000
 12        107    103.603       0.334
 13         12     13.272      -0.349
 14        110    106.736       0.316
 15         88     89.991      -0.210
 16       7712    154.648       0.000
```

This mover-stayer model fits very well. For the movers, those individuals who are susceptible to migrate, new place of residence does not depend on original residence. The fitted values for the diagonal (1.6, 95.2, 60.3, 154.6) are estimates of the numbers of movers in each category who did not happen to move in the period under observation. The number of stayers is obtained by subtracting these values from the observed diagonal values (116.4, 2031.8, 2487.7, 7557.4). We see that 92.6% (12193.3/13171) of the population is estimated as not being susceptible to migration.

Since the remaining two models are both based on the quasi- independence, but with additional parameters, we may expect that they will provide acceptable fits. The next model assumes both variables to have equal interval scales and fits the interaction between them.

```
MIGRANT BEHAVIOUR - FINGLETON (1984, P.142)

4. Uniform Association without Diagonal

scaled deviance = 4.3618 at cycle  3
           d.f. = 4         from 12 observations

Chi2 probability =  0.3594 for Chi2 =   4.362 with   4. d.f.

         estimate         s.e.      parameter
   1       0.4803       0.3582      1
   2        2.001       0.2119      M66(2)
   3        1.702       0.3613      M66(3)
   4        2.099       0.4099      M66(4)
   5        2.077       0.2220      M71(2)
   6        1.893       0.3581      M71(3)
   7        2.431       0.4050      M71(4)
   8     0.004893      0.07077      YY1_
   scale parameter taken as   1.000
```

The macro has created a new variable, YY1_, which is the product of the two linear scales. The parameter estimate gives the slope of the relationship between the two scales. Here, the zero slope reflects the independence of new residence from place of origin for the movers.

Finally, the row effects model is refitted, but now without the diagonal.

```
MIGRANT BEHAVIOUR - FINGLETON (1984, P.142)

 5. Row Effects Model without Diagonal

 scaled deviance = 1.5164 at cycle  3
            d.f. = 2       from 12 observations

 Chi2 probability =  0.4685 for Chi2 =   1.516 with   2. d.f.

          estimate          s.e.      parameter
     1   -0.05341         0.5039      1
     2      2.755         0.5297      M66(2)
     3      2.042         0.4808      M66(3)
     4      2.521         0.5024      M66(4)
     5      2.237         0.2602      M71(2)
     6      2.008         0.2879      M71(3)
     7      2.611         0.4889      M71(4)
     8     0.1708         0.2485      M66(1).ZZ1_
     9    -0.1437         0.2002      M66(2).ZZ1_
    10    0.02345         0.1710      M66(3).ZZ1_
    11      0.000       aliased       M66(4).ZZ1_
     scale parameter taken as   1.000
```

When the diagonal is eliminated, the slopes are greatly reduced as compared to model 2.

In both of these last models, the fit is very good, but too many parameters are included in the model. The mover-stayer model is retained as that best describing the data. The most important conclusions are the small proportion of movers in the population and the independence of arrival point from origin for these movers.

5. The Bradley-Terry Model

Occasionally, people may be asked to make a series of comparisons between pairs of objects, stating which is preferred. We have a square table showing how many individuals prefer each object as opposed to each other. The two variables are "prefer" and "not prefer", each with as many categories as there are objects to compare. The idea is to rank the objects in order of global preference for the group of people. If all people rank all objects, the rank is simply obtained from the number of positive preferences expressed (as in the ranking of teams in some sport). With unequal numbers, the problem is more complex.

Since we are concerned with ranking preferences, all ties may be ignored. Although we now have an incomplete table, with the diagonal missing, our problem is not resolved. We may construct a new table with one dimension being the object preferred and the other being the pair compared. For example, with four objects, we have

	1	2	3	4
(1,2)	n_{12}	n_{21}	-	-
(1,3)	n_{13}	-	n_{31}	-
(1,4)	n_{14}	-	-	n_{41}
(2,3)	-	n_{23}	n_{32}	-
(2,4)	-	n_{24}	-	n_{42}
(3,4)	-	-	n_{34}	n_{43}

This is now an incomplete table to which a model of quasi-independence may be fitted. We are testing to assure that general ranking of all objects does not depend on the specific pairs of comparisons made by the individual. If this model is acceptable, the parameters for the variable, object preferred (the columns of the table), give the rank.

This same model may also be developed in another way which does not require the table to be reconstructed. Instead, aspects of its symmetry are used. We construct a symmetric factor variable for the original table in the following way, here with six objects:

-	1	2	3	4	5
1	-	6	7	8	9
2	6	-	10	11	12
3	7	10	-	13	14
4	8	11	13	-	15
5	9	12	14	15	-

The factor variable has as many levels as there are possible paired combinations. Each symmetric pair has the same level. The diagonal will be eliminated by means of a zero weight, so that the values given to it are irrelevant, but must lie between 1 and the number of factor levels for the variable to be acceptable in $Fit. The reader may check that this new factor variable is identical to that for the rows of the reconstructed table above. The macro BTCT creates this symmetry variable and fits the model. We shall return to further applications of this symmetry variable in the next chapter.

Anderson (1980, p.357) provides a table of preferences expressed for a series of six collective facilities in a Danish municipality (Table 5.5). Unfortunately, he does not include information about which facilities are compared.

	Facility Number	Not Preferred Facility Number 1	2	3	4	5	6
	1	-	29	25	22	17	9
	2	49	-	35	34	16	14
Preferred	3	50	42	-	40	22	15
	4	54	43	37	-	33	16
	5	61	61	54	44	-	27
	6	69	64	63	62	51	-

Table 5.5 Preferences for Various Collective Facilities in Denmark (Andersen, 1980, p.357)

We shall apply the macro to this table:

```
PREFERENCE FOR COLLECTIVE FACILITIES IN DENMARK (ANDERSEN,1980,P.357)

 scaled deviance =   6.0721 at cycle  3
           d.f. = 10        from 30 observations

 Chi2 probability =  0.8102 for Chi2 =    6.072 with   10. d.f.

         estimate          s.e.       parameter
     1      3.404        0.1442       1
     2     0.4662        0.1454       PREF(2)
     3     0.6927        0.1464       PREF(3)
     4     0.8179        0.1465       PREF(4)
     5      1.434        0.1522       PREF(5)
     6      2.084        0.1652       PREF(6)
     7    -0.1844        0.1851       ZZ1_(2)
     8    -0.2563        0.1857       ZZ1_(3)
     9    -0.6944        0.1927       ZZ1_(4)
    10     -1.248        0.2057       ZZ1_(5)
    11    -0.3387        0.1757       ZZ1_(6)
    12    -0.4103        0.1770       ZZ1_(7)
    13    -0.8156        0.1862       ZZ1_(8)
    14     -1.312        0.2001       ZZ1_(9)
    15    -0.5102        0.1905       ZZ1_(10)
    16    -0.8963        0.1945       ZZ1_(11)
    17     -1.353        0.2036       ZZ1_(12)
    18    -0.9254        0.1936       ZZ1_(13)
    19     -1.379        0.2028       ZZ1_(14)
    20     -1.551        0.1996       ZZ1_(15)
    scale parameter taken as   1.000
```

The model fits very well. From the parameter estimates for PREF, the preferences are found to be ranked in the same order as they are presented in the table, with facility 6 most preferred.

6. Guttman Scales

A Guttman scale is constructed from a series of ordered yes/no questions such that once an individual replies yes (or no) to one question in the series, he/she should also reply yes (or no) to all subsequent questions. A typical series of questions to measure racial prejudice would be: 1) do you have immigrant friends? 2) would you buy a house next to an immigrant? 3) would you let your daughter marry an immigrant? The problem is that, most often, all people do not reply on the scale. The ordering is not respected and these individuals are unscalable.

With Q questions, a Guttman scale has Q+1 categories. We add another category, those who are not scalable. We then make the hypothesis that the responses to the Q questions for these unscalable individuals are independent. Note that unscalable individuals may fall on the scale by chance, in the same way as movers may stay put in the observed time interval. The sum of the probabilities for the Q+2 categories must be one; represent them by p_k (k=0, ... , Q+1). For the unscalable individuals, each question, i, has a probability, q_1^i, of reply yes ($q_2^i = 1-q_1^i$). Then, with Q = 3, we have

$$\begin{aligned} p_{111} &= p_1 + p_o q_1^1 q_1^2 q_1^3 \\ p_{112} &= p_2 + p_o q_1^1 q_1^2 q_2^3 \\ p_{122} &= p_3 + p_o q_1^1 q_2^2 q_2^3 \\ p_{222} &= p_4 + p_o q_2^1 q_2^2 q_2^3 \end{aligned} \qquad (5.4)$$

for individuals replying on the scale and

$$p_{jkl} = p_o q_j^1 q_k^2 q_l^3 \qquad (5.5)$$

for all responses not on the scale. As with the mover-stayer model, we eliminate those categories which are heterogeneous, i.e. contain both scalable and unscalable individuals - those on the scale by chance - by giving them zero weights and fit a quasi-independence model.

If the parameter estimates have been standardized to sum to zero (using the macro TRAN, for example), the probabilities of (5.5) are given by

$$q_{ji} = 1/(1 + \exp(2*\%PE)) \qquad (5.6)$$

where %PE is the parameter estimate supplied by GLIM. The probability, p_o, of being unscalable can now be directly calculated from (5.5) using any fitted value supplied by GLIM.

$$p_o = n_{jkl}/n_{...}/q_j^1/q_k^2/q_l^3 \qquad (5.7)$$

Since the p_{jkl} of (5.4) are simply the observed relative frequencies for these categories, the Q+1 probabilities of replying on the scale may be calculated by subtracting fitted from observed values.

Reply to Question				
1	2	3	4	Frequency
1	1	1	1	42
1	1	1	2	23
1	1	2	1	6
1	1	2	2	25
1	2	1	1	6
1	2	1	2	24
1	2	2	1	7
1	2	2	2	38
2	1	1	1	1
2	1	1	2	4
2	1	2	1	1
2	1	2	2	6
2	2	1	1	2
2	2	1	2	9
2	2	2	1	2
2	2	2	2	20

Table 5.6 Guttman Scale for Role Conflict (Fienberg, 1977, p.126)

The model will be applied to a scale of four questions indicating universalistic or particularistic values when confronted by four situations of role conflict (Table 5.6). We note the large numbers replying on the scale and the scattering of individuals off the scale, but also the combination (1212) which has a very large frequency.

The analysis is as follows:

```
ROLE CONFLICT (FIENBERG, 1977, P.126)

scaled deviance = 0.98849 at cycle  3
           d.f. = 6          from 11 observations

Chi2 probability =   0.9844 for Chi2 =   0.9885 with    6. d.f.

        estimate          s.e.        parameter
    1       1.788        0.1411       1
    2     -0.5919        0.1446       Q1
    3      0.2356        0.1757       Q2
    4      0.1164        0.1746       Q3
    5      0.7092        0.1534       Q4
    scale parameter taken as   1.000

Probability of replying by chance is   0.6826

Probabilities of replying yes to each question by chance are
  0.7656  0.3843  0.4421  0.1949

Probabilities of replying on the Guttman scale are
  0.1771  0.0350  0.0255  0.0314  0.0484
```

The model fits very well. However, the large number of persons replying off the scale, with 24 in the one specific way (1212), has already placed the original construction of the questions under suspicion. The results, indeed, indicate the probability of an unscalable reply to be 0.68. The probabilities of replying on the scale do not change regularly, and are all small, again placing the scale in question.

With data following a well-constructed Guttman scale, the probability of replying by chance should be very small so that the sum of probabilities of replying on the scale would be almost one. These probabilities would, then, indicate to which end of the scale individuals tend to lean.

CHAPTER 6

PATTERNS

1. Extremity Models

In this chapter, we study a number of special models applicable especially to square tables. In the previous chapter, we already encountered certain such models which applied to social mobility. The models of Markov chains in Chapter 2 also require square tables although in more than two dimensions. Here, we shall concentrate on certain patterns which may occur in such tables.

The simplest models for patterns may apply to any table, even if not square, since they only involve the symmetry of a very few cells, especially the corner cells. If the variables have an order, as is often the case in such tables, the corners are the extremes, hence the name of this class of models.

Consider a simple 2-way table relating opinions on whether or not grocery shopping is tiring to availability of a car (Table 6.1).

	Grocery Shopping is Tiring				
Availability of a Car	Disagree	Tend to Disagree	In Between	Tend to Agree	Agree
No Car Available	55	11	16	17	100
Sometimes Car Available	101	7	18	23	103
Car Always Available	91	20	25	16	77

Table 6.1 Oxford Shopping Survey (Fingleton, 1984, p.10)

We first fit the independence model to verify if there is a relationship between the two variables.

```
OXFORD SHOPPING SURVEY (FINGLETON, 1984, P.10)

Independence Model

scaled deviance = 23.871 at cycle  3
          d.f. =  8

Chi2 probability =  0.0025 for Chi2 =   23.87 with   8. d.f.
```

	estimate	s.e.	parameter
1	4.281	0.08719	1
2	-1.872	0.1739	TIRE(2)
3	-1.432	0.1449	TIRE(3)
4	-1.484	0.1480	TIRE(4)
5	0.1254	0.08729	TIRE(5)
6	0.2361	0.09482	CAR(2)
7	0.1404	0.09689	CAR(3)

scale parameter taken as 1.000

unit	observed	fitted	residual
1	55	72.284	-2.033
2	11	11.121	-0.036
3	16	17.266	-0.305
4	17	16.388	0.151
5	100	81.941	1.995
6	101	91.535	0.989
7	7	14.082	-1.887
8	18	21.865	-0.827
9	23	20.753	0.493
10	103	103.765	-0.075
11	91	83.181	0.857
12	20	12.797	2.013
13	25	19.869	1.151
14	16	18.859	-0.658
15	77	94.294	-1.781

OXFORD SHOPPING SURVEY (FINGLETON, 1984, P.10)

Poisson Residuals

Score Test Coefficient of Sensitivity

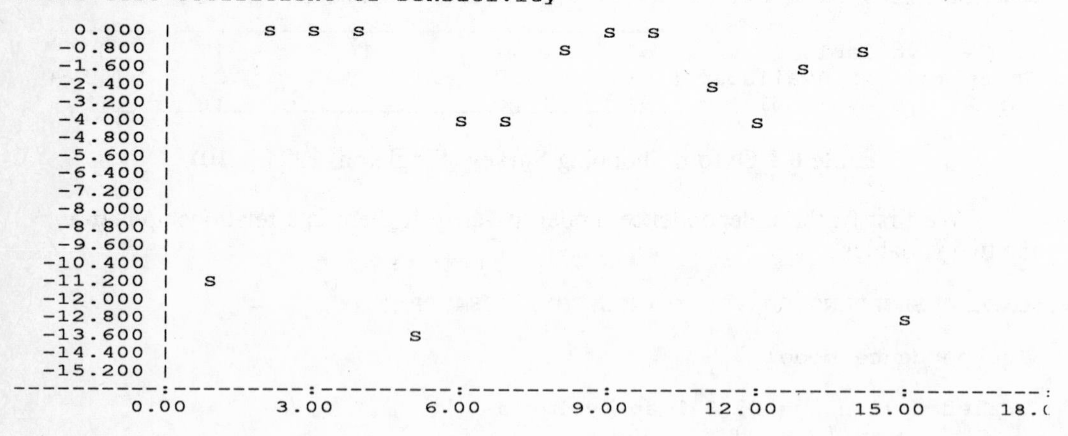

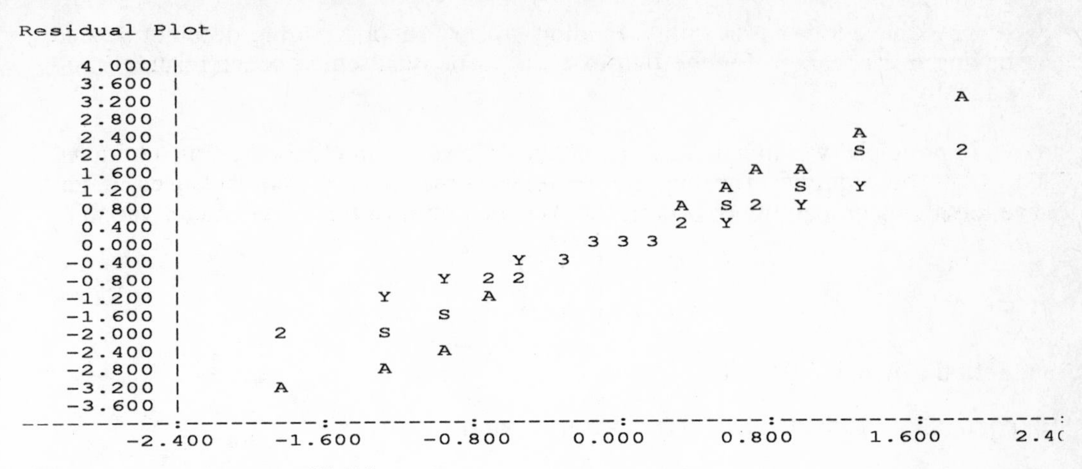

This model is rejected. From the score test coefficient of sensitivity, we see that two of the observations fitting most poorly are the two extremes, those in the upper left and lower right corners. One possibility is that these two conflictual extremes, disagree that shopping is tiring when no car is available and agree that shopping is tiring when a car always available, are the exceptions to independence, since they should have a lower probability of occurring.

We shall construct a factor variable with two levels, contrasting these extremes to the rest of the table:

```
2 1 1 1 1
1 1 1 1 1
1 1 1 1 2
```

and add this to the model.

```
OXFORD SHOPPING SURVEY (FINGLETON, 1984, P.10)

Extreme Ends Model

scaled deviance = 10.205 (change =   -13.67) at cycle  3
           d.f. =  7      (change =    -1    )

Chi2 probability =  0.1763 for Chi2 =   10.21 with   7. d.f.

         estimate        s.e.       parameter
    1       4.425      0.09761      1
    2      -1.978       0.1761      TIRE(2)
    3      -1.538       0.1473      TIRE(3)
    4      -1.590       0.1504      TIRE(4)
    5      0.1453      0.08831      TIRE(5)
    6      0.1064       0.1001      CAR(2)
    7      0.1623      0.09803      CAR(3)
    8     -0.4007       0.1102      EX2(2)
    scale parameter taken as   1.000
```

The model now fits very well and the score test coefficients (not shown) no longer indicate a problem with the extremes. The parameter value (-0.4007) confirms that the two corners have lower probability. Finding grocery shopping tiring does not depend on having a car, except for the two extreme responses, which occur relatively too infrequently.

In principle, we do not need to continue. However, in other cases, an additional step might be required. The opposite extreme corners, the concordant ones, might have too high a probability of occurrence. We then set up a three-level factor variable:

```
2 1 1 1 3
1 1 1 1 1
3 1 1 1 2
```

and refit the model to the data.

```
OXFORD SHOPPING SURVEY (FINGLETON, 1984, P.10)

Four Corners Model

scaled deviance = 9.6469 (change = -0.5585) at cycle  3
           d.f. = 6      (change = -1       )

Chi2 probability =  0.1393 for Chi2 =    9.647 with   6. d.f.

           estimate        s.e.      parameter
    1         4.556      0.2023      1
    2        -2.072      0.2184      TIRE(2)
    3        -1.632      0.1959      TIRE(3)
    4        -1.684      0.1982      TIRE(4)
    5        0.1439     0.08823      TIRE(5)
    6     -0.005965      0.1820      CAR(2)
    7        0.1608     0.09794      CAR(3)
    8       -0.5307      0.2070      EX4(2)
    9       -0.1497      0.2012      EX4(3)
    scale parameter taken as   1.000
```

For our present example, the model must fit well since the two corner model did. No significant improvement occurs. Surprisingly, the parameter value (-0.1497) indicates that the two concordant corners also have lower probability than the body of the table, although this time the difference is not significant.

If a single cell is an extreme case, the easiest way to account for it in a model with GLIM is by giving it a zero weight. Paradoxically, this is equivalent to creating a two-level factor variable, where only that one cell has the second factor level, but gives a neater result.

2. Symmetry Models

The remaining models in this chapter apply only to square tables. All exploit the symmetry of such tables; in fact, a near relative of them was already encountered in the preceding chapter, the Bradley-Terry model. All of these models, except one, are provided by a single macro, SYCT. We present them in three sections, using three different examples. Only the relevant parts of the output of the macro will be presented for each example.

A completely symmetrical table is one in which the probabilities in opposing cells across the diagonal are equal:

$$p_{ik} = p_{ki} \tag{6.1}$$

The corresponding log linear model is

$$\log(F_{ik}) = \alpha_{ik} \quad \text{with } \alpha_{ik} = \alpha_{ki} \tag{6.2}$$

Note that mean parameters for the margins are not fitted. The factor variable (ZZ1_) used to fit this model is that used in the Bradley-Terry model:

-	1	2	3	4	5
1	-	6	7	8	9
2	6	-	10	11	12
3	7	10	-	13	14
4	8	11	13	-	15
5	9	12	14	15	-

with weight zero for the diagonal. However, the Bradley-Terry model contained one mean parameter, that for preferences.

1985	1981 PS	PRL	PSC	Ecolo	PCB	BB
Socialist (PS)	281	14	9	16	4	4
Liberal (PRL)	12	164	13	4	1	6
Social-Christian (PSC)	5	10	121	8	1	1
Ecology (Ecolo)	6	0	1	50	0	1
Communist (PCB)	1	0	0	2	14	0
Blank Ballot (BB)	2	1	0	0	0	11

Table 6.2 Voting Changes between Belgian Elections, 1981-1985 (R. Doutrelepont)

We shall fit this model to data on how voters, interviewed outside the polling station in the October 1985 Belgian election, stated they had just voted and how they had voted in the previous election (Table 6.2). This is a retrospective study, rather than a panel.

The relevant sections of the output from the macro are as follows:

```
BELGIAN ELECTIONS - 1981-1985 - VOTING CHANGES

2. Symmetry Model

scaled deviance = 33.054 at cycle  7
           d.f. = 15       from 30 observations

Chi2 probability =  0.0047 for Chi2 =   33.05 with   15. d.f.
```

```
        estimate          s.e.        parameter
  1        2.565        0.1961        1
  2      -0.6190        0.3315        ZZ1_(2)
  3      -0.1671        0.2897        ZZ1_(3)
  4       -1.649        0.4883        ZZ1_(4)
  5       -1.466        0.4529        ZZ1_(5)
  6      -0.1226        0.2863        ZZ1_(6)
  7       -1.872        0.5371        ZZ1_(7)
  8       -3.258         1.019        ZZ1_(8)
  9       -1.312        0.4258        ZZ1_(9)
 10       -1.061        0.3867        ZZ1_(10)
 11       -3.258         1.019        ZZ1_(11)
 12       -3.258         1.019        ZZ1_(12)
 13       -2.565        0.7338        ZZ1_(13)
 14       -3.258         1.019        ZZ1_(14)
 15       -10.26         20.09        ZZ1_(15)
 scale parameter taken as    1.000
```

Since the model is rejected, the probability of changing vote in either direction between each pair of parties is not the same.

A weaker hypothesis is that of quasi-symmetry: the table would be symmetric if it were not for the distorting effect of the marginal totals. In our example, this is the effect of a changing proportion of votes received by the different parties between the two elections. We simply add the two mean parameters to the model (6.2). (In a sense, the Bradley-Terry model lies between symmetry and quasi-symmetry, since it contains one set of mean parameters.)

```
BELGIAN ELECTIONS - 1981-1985 - VOTING CHANGES

3. Quasi-symmetry Model

scaled deviance = 10.101 (change =   -22.95) at cycle  9
           d.f. = 10     (change =    -5   ) from 30 observations

Chi2 probability =  0.4319 for Chi2 =   10.10 with   10. d.f.

        estimate          s.e.        parameter
  1        2.587        0.2496        1
  2      -0.7439        0.3763        ZZ1_(2)
  3       -1.189        0.4546        ZZ1_(3)
  4       -2.353        0.7509        ZZ1_(4)
  5       -2.510        0.7065        ZZ1_(5)
  6      -0.2275        0.3445        ZZ1_(6)
  7       -2.886        0.6379        ZZ1_(7)
  8       -3.951         1.167        ZZ1_(8)
  9       -2.348        0.6932        ZZ1_(9)
 10       -2.122        0.5214        ZZ1_(10)
 11       -4.016         1.158        ZZ1_(11)
 12       -4.340         1.148        ZZ1_(12)
 13       -3.894        0.8486        ZZ1_(13)
 14       -4.783         1.097        ZZ1_(14)
 15       -13.59         54.27        ZZ1_(15)
 16     -0.04484        0.3158        V81(2)
 17       0.1959        0.3554        V81(3)
 18        1.490        0.4259        V81(4)
 19        1.084        0.7559        V81(5)
```

```
20        1.516      0.6797       V81(6)
21        0.000     aliased       V85(2)
22        0.000     aliased       V85(3)
23        0.000     aliased       V85(4)
24        0.000     aliased       V85(5)
25        0.000     aliased       V85(6)
 scale parameter taken as   1.000
```

This model fits the data very well. The probability of shifting in either direction between each pair of parties is the same after taking into account the overall change in voting behaviour between the two elections. Change between Socialist and Liberal is most probable (all other estimates are negative with regard to it) and between Communist and blank ballot least (there are none).

A further model, marginal homogeneity, is closely related to the previous two. Suppose the marginal totals are symmetric but the body of the table is not. The distribution of votes at the two elections is identical but the probability of shift between each pair of parties is not the same in both directions. Marginal homogeneity plus quasi- symmetry equals symmetry; the Chi-squares of these models obey this equation. Symmetry obviously implies marginal homogeneity. A special macro, MHCT, is required for this model.

```
BELGIAN ELECTIONS - 1981-1985 - VOTING CHANGES

Marginal Homogeneity Model

scaled deviance =    23. at cycle    10.
      d.f. =     5.
   (no convergence yet)

Chi2 probability =  0.00 for Chi2 =    23. with   5. d.f.

      estimate       s.e.      parameter
  1      -0.59        1.5      C1_
  2      -0.66        1.5      C2_
  3      -0.55        1.7      C3_
  4      0.032        1.7      C4_
  5     -0.063        2.5      C5_
  6        0.0    aliased      C6_
  7        0.0    aliased      C7_
  8        0.0    aliased      C8_
  9        0.0    aliased      C9_
  scale parameter taken as    33.
```

As might be expected, this model is not acceptable for these data, since symmetry was not, while quasi-symmetry was.

We may note that marginal homogeneity is not a log linear model (hence, the special macro), the third we have encountered in this book; the first two were the log-multiplicative and proportional odds models.

In terms of Markov chains, quasi-symmetry is known as reversibility, since the same proportion of individuals is changing position in each direction, while marginal homogeneity is the equilibrium state, since the margins are not changing over time.

3. Diagonal Models

Several models take into account the diagonal symmetry of square tables. A first model bears some similarity to the mover-stayer model. We differentiate those who do not change from those who do, i.e. the diagonal from the rest. However, in distinction to the mover-stayer model, we here consider the diagonal members to be homogeneous. This is the main diagonal or loyalty model, since, in voting behaviour, we are distinguishing those who are loyal to a party (those on the diagonal) from those who are not. The factor variable (ZZ2_) is

```
2 1 1
1 2 1
1 1 2
```

We apply this model to the two 1974 British elections (Table 6.3).

February	October Conservative	Liberal	Labour
Conservative	170	20	3
Liberal	22	70	28
Labour	6	12	227

Table 6.3 Changes in Vote between the Two British Elections of 1974 (Fingleton, 1984, p.131)

The macro gives the following results:

```
BRITISH ELECTION VOTE 1974 (FINGLETON, 1984, P.131)

5. Main Diagonal (Loyalty) Model

scaled deviance = 53.175 at cycle  4
           d.f. =  3

Chi2 probability =  0.0000 for Chi2 =   53.17 with   3. d.f.

         estimate          s.e.       parameter
    1       2.791        0.1251       1
    2     -0.6495        0.1755       OCT(2)
    3      0.2078        0.1477       OCT(3)
    4    -0.01844        0.1694       FEB(2)
    5     0.07539        0.1506       FEB(3)
    6       2.312        0.1186       ZZ2_(2)
    scale parameter taken as   1.000
```

Although the Chi-square is greatly reduced from that for the independence model (613.1 with 4 d.f.), the model is not satisfactory. Loyalty is an important factor, especially with two so closely spaced elections, but it is not a sufficient explanation of the pattern in the data.

We now take into consideration the idea that the parties may be ordered and that changing vote by one step in either direction on the scale has a different probability than that for two steps, and so on, for greater distances. The factor variable (ZZ3_) is now

```
1 2 3
2 1 2
3 2 1
```

the symmetric minor diagonal model.

```
6. Symmetric Minor Diagonal Model

scaled deviance = 4.0486 (change =  -49.13) at cycle  3
         d.f. = 2        (change =   -1   )

Chi2 probability =  0.1321 for Chi2 =   4.049 with   2. d.f.

        estimate         s.e.       parameter
    1      5.148      0.07518       1
    2    -0.6112       0.1655       OCT(2)
    3     0.2496       0.1897       OCT(3)
    4    -0.2885       0.1663       FEB(2)
    5    0.01777       0.1935       FEB(3)
    6     -1.773       0.1305       ZZ3_(2)
    7     -3.784       0.3372       ZZ3_(3)
    scale parameter taken as   1.000
```

This model fits the data very well. The parameter estimates of ZZ3_ (0.000, -1.773, -3.784) indicate that the probability of changing vote decreases steeply with distance between the parties, in the order presented in the table.

If the model did not fit, we could extend it further and take steps with different probabilities in each direction, the asymmetric minor diagonal model, with factor variable (ZZ4_)

```
1 2 3
4 1 2
5 4 1
```

As expected, for these data, the model fits well, since the preceding one did.

```
BRITISH ELECTION VOTE 1974 (FINGLETON, 1984, P.131)

7. Asymmetric Minor Diagonal Model

scaled deviance = 1.4032 (change =  -2.645) at cycle  3
         d.f. = 1        (change =  -1     )

Chi2 probability =  0.2362 for Chi2 =   1.403 with   1. d.f.

        estimate         s.e.       parameter
    1      5.151      0.07513       1
    2      1.081       0.2427       OCT(2)
    3      3.622       0.4136       OCT(3)
    4     -1.983       0.2478       FEB(2)
    5     -3.359       0.4151       FEB(3)
    6     -3.371       0.2637       ZZ4_(2)
```

```
7        -7.674       0.7143       ZZ4_(3)
8       -0.1977       0.2737       ZZ4_(4)
9         0.000      aliased       ZZ4_(5)
scale parameter taken as   1.000
```

With only three parties, not all parameters can be estimated, hence the alias.

Another possibility is to combine the minor diagonals model with symmetry. When this model fits, it indicates that we would have symmetry if it were not for the unequal probabilities of larger and smaller steps. In this case, a full factor variable is not necessary because of the effect of the symmetry variable in the model. It is sufficient to define a variable (YY1_)

```
1 2 3
1 1 2
1 1 1
```

and fit it with the symmetry factor variable, but with no mean parameters.

```
4. Minor Diagonals-Symmetry Model

 scaled deviance = 4.2759 at cycle  3
            d.f. = 1       from 6 observations

 Chi2 probability =  0.0387 for Chi2 =   4.276 with   1. d.f.

          estimate          s.e.      parameter
   1         2.857        0.2025      1
   2        0.3448        0.2241      YY1_(2)
   3       -0.6931        0.7071      YY1_(3)
   4        -1.066        0.4557      ZZ1_(2)
   5      -0.04879        0.2209      ZZ1_(3)
   scale parameter taken as   1.000
```

As in Section 2 above, ZZ1_ is the symmetry variable; here, YY1_ is the new minor diagonal variable. (The macro, SYCT, prints out a list of the values of all variables created at the end of its output, so that the user can verify to what each variable corresponds.)

This model is the similar to the symmetric minor diagonal model, but without the margins fixed. The fit is reasonably good, but we retain the symmetric minor diagonals model, since it fits better and has more degrees of freedom. It would appear that the symmetric minor diagonals model has more parameters, but here the diagonal of the table has been eliminated for symmetry.

4. Distance and Loyalty Models

Diagonal models assume an equal distance among all adjacent pairs of categories. Distance models relax this assumption to allow different intervals among the categories. A distinct variable is introduced for each adjacent interval, i.e. K-1 variables for K categories. A model with these variables plus the two mean variables may be called a pure distance model. For a 4x4 table, the series of variables (C1_,

C2_, ...) is

```
2 1 1 1        2 2 1 1        2 2 2 1
1 2 2 2        2 2 1 1        2 2 2 1
1 2 2 2        1 1 2 2        2 2 2 1
1 2 2 2        1 1 2 2        1 1 1 2
```

In the macro, SYCT, these are modified to sum to zero, as if by the macro, TRAN.

We fit the model to voting changes between the 1966 and 1970 British elections (Table 6.4).

1966	**1970** Conservative	Liberal	Labour	Abstention
Conservative	68	1	1	7
Liberal	12	60	5	10
Labour	12	3	13	2
Abstention	8	2	3	6

Table 6.4 Voting Changes between 1966 and 1970 British Elections (Upton, 1978, p.119)

The results which interest us from the macro are given below:

```
BRITISH ELECTION VOTE 1966 AND 1970 (UPTON, 1978, P.119)

8. Pure Distance Model

scaled deviance = 64.230 at cycle  5
           d.f. =  6

Chi2 probability =   0.0000 for Chi2 =    64.23 with    6. d.f.

         estimate         s.e.      parameter
   1        3.099       0.1928      1
   2      -0.8883       0.2172      V70(2)
   3       -1.839       0.2916      V70(3)
   4       -1.601       0.3273      V70(4)
   5       0.4428       0.2129      V66(2)
   6      -0.4539       0.2718      V66(3)
   7      -0.8267       0.3273      V66(4)
   8      -0.7572       0.1031      C1_
   9      -0.2691      0.09916      C2_
  10     -0.09417       0.1473      C3_
  11        0.000      aliased      C4_
  12        0.000      aliased      C5_
  13        0.000      aliased      C6_
  14        0.000      aliased      C7_
  15        0.000      aliased      C8_
  16        0.000      aliased      C9_
    scale parameter taken as   1.000

  unit  observed     fitted   residual
     1        68     68.000      0.000
     2         1      6.153     -2.077
     3         1      1.388     -0.329
```

4	7	1.459	4.588
5	12	23.290	-2.339
6	60	43.557	2.491
7	5	9.827	-1.540
8	10	10.326	-0.101
9	12	5.546	2.741
10	3	10.372	-2.289
11	13	6.867	2.341
12	2	7.215	-1.942
13	8	3.164	2.718
14	2	5.918	-1.611
15	3	3.918	-0.464
16	6	6.000	0.000

BRITISH ELECTION VOTE 1966 AND 1970 (UPTON, 1978, P.119)

Poisson Residuals

Score Test Coefficient of Sensitivity

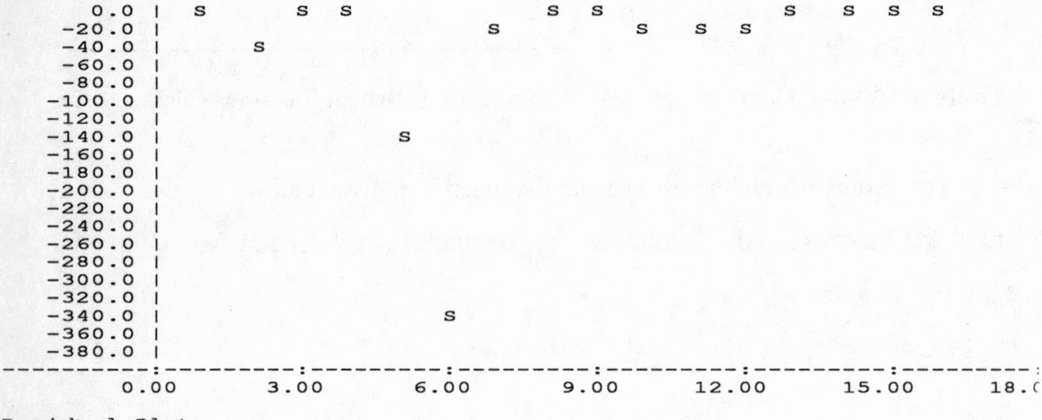

Residual Plot

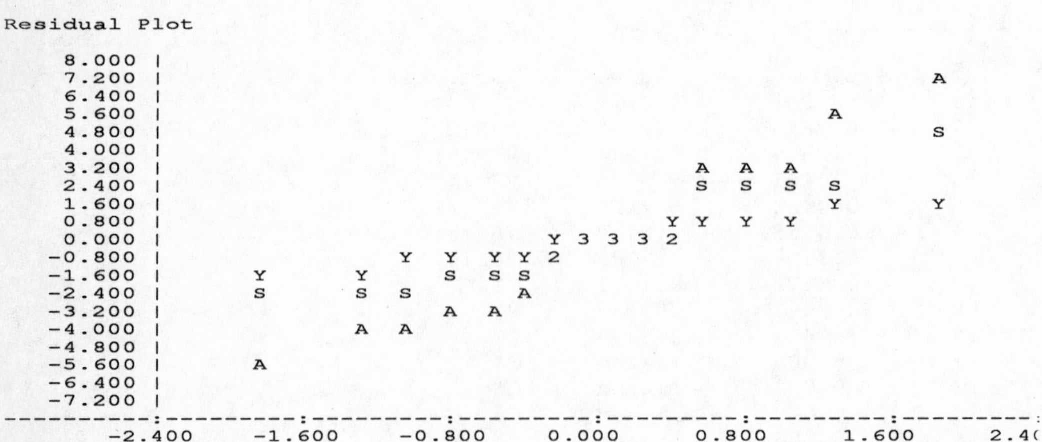

Points Y represent 45 line

The model is rejected. However, the problem with such a model for voting data is that it does not take into account party loyalty. If we inspect the residual table and the score test coefficient, we see that this is only important for the interior diagonal elements (2 and 3), and not for the two extremes. The stability of the Liberal vote is particularly under-estimated.

We add the loyalty variable (ZZ2_), the main diagonal factor of the previous section, to obtain the loyalty-distance model.

```
BRITISH ELECTION VOTE 1966 AND 1970 (UPTON, 1978, P.119)

9. Loyalty-Distance Model

scaled deviance = 6.0702 (change =  -58.16) at cycle  4
          d.f. = 5       (change =   -1   )

Chi2 probability =  0.2989 for Chi2 =   6.070 with   5. d.f.

         estimate        s.e.      parameter
    1     2.331         0.2340     1
    2    -1.063         0.2632     V70(2)
    3    -1.560         0.2996     V70(3)
    4    -1.505         0.3109     V70(4)
    5     0.9108        0.2584     V66(2)
    6     0.02230       0.2866     V66(3)
    7    -0.9232        0.3109     V66(4)
    8     0.06429       0.1541     C1_
    9     0.3150        0.1577     C2_
   10     0.4398        0.1786     C3_
   11     0.000        aliased     C4_
   12     0.000        aliased     C5_
   13     0.000        aliased     C6_
   14     0.000        aliased     C7_
   15     0.000        aliased     C8_
   16     0.000        aliased     C9_
   17     2.708         0.3966     ZZ2_(2)
    scale parameter taken as  1.000
```

The model now fits very well. (In fact, the symmetric minor diagonal model also fits these data well; it too takes party loyalty into account.) On the ordered party scale, Conservative and Liberal are closest neighbours (0.0643) and Labour and abstention are most distant (0.4398).

Another possibility to accommodate the inflated main diagonal elements is to combine the pure distance model with the mover-stayer model to give a distance model without main diagonal.

```
BRITISH ELECTION VOTE 1966 AND 1970 (UPTON, 1978, P.119)

10. Distance without Main Diagonal

scaled deviance = 4.2995 at cycle  4
          d.f. = 4       from 12 observations

Chi2 probability =  0.3673 for Chi2 =   4.300 with   4. d.f.

         estimate        s.e.      parameter
    1     2.015         0.4357     1
    2    -1.473         0.4581     V70(2)
    3    -1.348         0.4196     V70(3)
    4    -0.6278        0.3413     V70(4)
    5     0.6938        0.4075     V66(2)
```

```
 6      0.1263       0.4582     V66(3)
 7     -0.02927      0.4937     V66(4)
 8      0.000       aliased     C1_
 9      0.2525       0.1527     C2_
10      0.000       aliased     C3_
11      0.000       aliased     C4_
12      0.000       aliased     C5_
13      0.000       aliased     C6_
14      0.000       aliased     C7_
15      0.000       aliased     C8_
16      0.000       aliased     C9_
 scale parameter taken as   1.000
```

This model also fits the data very well, but with one less degree of freedom than the previous model. Whereas the main diagonal variable gives a constant factor level to the whole diagonal, eliminating the diagonal is equivalent to giving each category of the diagonal a different level. This is evidently unnecessary for these data.

Several remarks should be made in conclusion. All of the models of this chapter are suitable for the study of social mobility tables, already introduced in the previous chapter. However, care must be taken with small tables, especially 3x3, since in this case, a number of the models are identical. We already noted the problem of aliased parameter estimates in one model of Section 3 above. Note, also, that all of these models may be relatively easily extended to multi-dimensional tables covering more than two time points, although the supplied macros no longer apply, and suitable variables must be constructed.

APPENDIX I

GLIM COMMANDS

The following table is adapted to GLIM 3.77 from the Hull University GLIM card for GLIM 3.12 published in the GLIM Newsletter Number 3 (1980).

```
Implementation details:
 Site                        Liege
 Machine                     Amiga
 Operating system            AmigaDOS
 Mark of GLIM                3.77 Update 2
 O.S. GLIM entry command     GLIM

Special symbols ($ENV I):
 Directive                   $
 Repetition                  :
 Function                    %
 Substitution                #
 End of record               !
 Quote (Text)                '
 Separator                   ;
 Output request              []
 Greater than                >
 Less than                   <
 And                         &
 Or                          ?
 Not                         /
 Query                       ?
 Modulus                     |
 Largest Integer 2147483639

Input/Output channel numbers:
 Default input (keyboard)     9
 Default output (screen)      9
 Secondary input (data)       1
 Secondary input (program)    5
 Secondary output (listing)   6

Full character set:
 A ... Z  a ... z  0 ... 9 underline
 space newline comma  special symbols
 operators: + - * / **
 brackets: ( )

Names:
 Not more than 4 characters of a
 name are significant. The directive
```

```
Directives:
  In a description of a directive,
  "int" specifies an integer value and
  "number" a value that may contain a
  sign and a decimal point. "scalar"
  must be a scalar identifier but "id"
  may be a scalar or variate identifier
  in context. "macro" is a macro
  identifier. "option-list" is
  directive specific keywords. Items in
  [ ] are optional.
$ACcuracy int       No. of digits for
                    output
$ALias              Switch to include/
                    exclude intrinsically
                    aliased parameters
$Argument macro items
                    Define up to 9
                    arguments for macro.
                    Item may be name,
                    %int, %scalar, or *
$ASSign vector1 = id [,id]s
                    Concatenate list of
                    values.
$CAlculate expression
                    Evaluate and
                    optionally print
                    value
$Comment string    Non-executable
                    text
$CYcle [int1[int2[number1[number2]]]
                    No. of cycles and
                    printing frequency
$DAta [length] id's Define names for
                    $Read or $DINput
$DElete id's
$DINput channel [width]
                    Read data from file
$Display letters    Use after fitting
                    model. Letters: A C
                    D E L M R S T U V W
$                   Dummy directive
$DUmp [channel]     Save current state
$ECho               Switch to print
                    back input
```

symbol is the first significant character of a directive. Directives may therefore be type in full, as just the symbol plus 3 characters, or they may be further shortened to the portion capitalized in the list of directives. Lower case letters are interpreted as the upper case equivalent.

These names are system defined:

Scalars:

%A %B ... %Z Ordinary scalars
%JN Job number. Incremented by $End
%NU No. of UNITS
%DV Scaled deviance after fit
%DF Degrees of freedom after first cycle
%X2 Generalized Pearson Chi-square after each cycle
%SC SCALE or mean deviance
%CL Program control stack level
%ML No. of elements in (co)variance matrix of parameters. Length of vector %VC
%PL No. of non-intrinsically aliased parameters. Length of vector %PE
%PI Pi to machine accuracy
%HEL 1 if $Help on, else 0
%ECH 1 if $ECho on, else 0
%WAR 1 if $WArning on, else 0
%VER 1 if $VErify on, else 0
%PAG 1 if $PAGe on, else 0
%PIC primary input channel number
%PIL record length of prim. inp. ch.
%CIC current input channel
%CIL record length of curr. inp. ch.
%POC primary output channel number
%POL record length of prim. out. ch.
%POH height of prim. out. channel
%COC current output channel number
%COL record length of curr. out. ch.
%COH height of curr. out. channel
%PDC primary dump channel number
%PLC primary library channel number
%ACC accuracy setting
%IM 1 if GLIM in interactive mode
%TRA transcription code
%ERR error distribution code
%LIN link code
%YVF 1 if y-variate specified, else 0
%BDF 1 if binomial denom. specified
%PWF 1 if prior weight specified
%OSF 1 if offset specified, else 0
%A1...%A9 1 if nth arg. of macro set
%Z1...%Z9 scalars reserved for macros
%CYC maximum number of cycles
%PRT printing frequency

$EDit [int1[int2]] vectors numbers
$End End of job. Clears user space. Does not reset pseudo-random numbers
$Endmac End of macro definition
$ENVironment [channel] letters Letters: C D E G I P R S U : Channels/ Direct./ Pass/ Graphics/Imp./PCS/ Random Seeds/System struct./Usage
$ERror letter[id] Letters: B G N P Binomial (needs id)/ Gamma/Normal/Poisson
$EXit [int] Pop program control stack 'int' levels. See $SKip
$EXTract id's Assign values from SSP to identifiers %VC, %PE, or %VL
$FActor [length][id level]s
$FINish End of file marker after subfiles. May cause file to be rewound
$Fit [model formula]
$FOrmat FREE or FORTRAN format
$GRAph not implemented
$GROup [vector2 =] vector1 [Values vector4][Interval [*] vector3 [*]] Regroup values in vector1 with vector3 as domain and vector2 as range
$Help Switch to give extended error messages
$HIstogram [option-list][vector1 [/vector2]]s ['string'[vector3]] Plot histogram(s) vector1, with weight from vector2 for each factor level of vector3
$Input channel [width] [subfiles]
$LAyout Not implemented
$LInk letter [number] Declare link function. Letters: C E G I L P R S : Comp. log-log/Exp. (needs no.)/Logit/Identity/ Log/Probit/Recip./ Square root
$Look [option-list] vectors or scalars
$LSeed [in1[in2[in3]]]

```
 %CC  convergence criterion
 %TOL aliasing tolerance
 %S1 %S2 %S3 seeds for random num.gen.

Vectors (length in brackets):
 %FV  Fitted values (%NU)
 %LP  Linear predictors (%NU)
 %WT  Iterative weights (%NU)
 %WV  Working vector for iterative
       models (%NU)
 %YV  Dependent variate (Y) values
       (%NU)
 %BD  Binomial denominator (%NU)
 %PW  Prior weights (%NU)
 %OS  Offset (%NU)
 %DR  Derivative d(eta)/d(mu) (%NU)
 %VA  Variance function values (%NU)
 %DI  Deviance increment (%NU)
 %GM  Grand mean used in FITs (%NU)
 %VC  Non-intrinsically aliased
       parameter (co)variance
       matrix (%ML)
 %PE  Non-intrinsically aliased
       parameter estimates (%PL)
 %VL  Variances of linear predictors
       (%NU)
 %RE  Weights for Plotting or DISplay
       (%NU)

Functions:
 X is a variate or scalar, depending
  upon context and k, n integer
  scalars
 %ANG(X)   arcsin(sqrt(X))
 %EXP(X)   e**X
 %LOG(X)   ln(X) base e
 %SIN(X)   sin(X)
 %SQRt(X)  square root
 %NP(X)    Normal probability integral
            infinity to X
 %ND(X)    Normal deviate, inverse of
            %NP  0<X<1
 %TR(X)    Integer X, truncated toward
            0
 %GL(k,n)  Factor levels 1 to k in
            blocks of n
 %CU(X)    Cumulative sums of X
 %SR(0)    Pseudo-random real on [0,1]
 %SR(n)    Pseudo-random integer on
            [0,n]
 %LR(0)    Pseudo-random real on [0,1]
 %LR(n)    Pseudo-random integer on
            [0,n]

Logical operators:
  <  <=  =  /=  >=  >
 Dyadic  AND   &
 Dyadic  OR    ?
 Monadic NOT   /
```

```
$Macro macro space string $Endmac
$MANual            Not implemented
$MAP [vector2=] vector1 [Values
  vector4] [Interval [*]vector3[*]]
                   values of vector1
                   mapped with vector3
                   as domain and vector2
                   as range
$Offset [id]       Declare a priori
                   known component in
                   fit
$OUtput [channel[width[height]]
                   If channel=0,
                   switches off output
$OWn macro1 macro2 macro3 macro4
                   macro1: produce %FV
                   from %LP
                   macro2: produce %DR
                   macro3: produce %VA
                   macro4: produce %DI
$PAGe              Switch to pause
                   output
$PASs              Not implemented
$PAUse             Open a new multi-
                   tasking Command
                   Line window
$Plot [option-list] yvectors xvector
  ['string'[vector]]
                   Up to 9 yvectors.
                   Vector specifies
                   factor levels
$PRint [option-list] [item]s
                   Item is identifier.
                   string, *int or /
$Read numbers      Read values to id's
                   named in $DAta
$RECycle[int1[int2][number1[number2]]]
                   As cycle, but starts
                   with %FV
$REInput channel  [width] [subfiles]
$REStore [channel] Restart from DUMP
$RETurn            Pop input channel
                   stack by 1 level
$REwind [channel]
$SCale [number]    If number>0, use as
                   scale factor; else
                   estimate scale
$SET option        Specify batch or
                   interactive mode
$SKip int          Pop program counter
                   stack 'int' levels
                   unless in $WHile
$Sort vector1 [vector2 or int2[vector3
  or int3]]        Sort vector2 into
                   vector1 based on
                   vector3. Use int2 for
                   ranks and int3 for
                   circular lags
```

%LT %LE %EQ %NE %GE %GT
These take 2 arguments: e.g. %LT(X,Y)
TRUE=1. FALSE=0.
%IF(conditional expression,X1,X2)
Returns X1 if true, X2 if false.
Logical values may be combined:

```
   %op1 AND %op2  by %op1*%op2
   NOT %op1       by 1-%op1
   %op1 NOR %op2  by %EQ(%op1+%op2,0)
   %op1 OR %op2   by %NE(%op1+%op2,0)
   %op1 EOR %op2  by %NE(%op1,%op2)
```

Formal arguments:
%1 ... %9
%scalar e.g. %%A

Operators in precedence order:
1. functions, monadic operators,
2. **
3. * and dyadic /
4. dyadic + and dyadic -
5. relational operators
6. &
7. ?
8. =

Layout:
Items must be separated by space or newline. $SUbfile or $FINish must be the first directive on any line in which they occur. Text following $End, $FINish or $RETurn on the same line is ignored.

$SSeed [int1[int2[int3]]]
$STop — End of session
$SUBfile id space text $RETurn — External object
$SUSpend — Temporary reversion to primary input
$SWitch scalar macros — Conditionally execute macro from list
$Tabulate [option-list] [THE (vector1 or *) statistic [number]] [WITH vector2 or *] [FOR (vector3 [; vector4]s) or *] [INTO (vector5) or output-request or *][USING (vector6) or output-request or *] [BY (vector7 [; vector8]s) or (scalar1 [; scalar2]s) or *]
where output-request is []. For each FOR vector, the statistic weighted by WITH vector2 is calculated from the THE vector1 output classification is stored in BY vectors or scalars, the resultant weight in USING vector6 and the calculated statistic values in the INTO vector5. Output-request prints a table. Statistic may be Mean, Total, Variance, Deviation, Smallest, Largest, Fifty, Percentile, Interpolate.
$TPrint [option-list] vector1 [; vector2]s [((vector3 [; vector4]s) or (number1 [; number2])) or *]
Print values of vector1 and vector2 as body of table classified by vector3 and vector4
$TRanscript [Input] [Verify] [Warn] [Fault] [Help] [Ordinary]
Specify what is written to transcript file (Must have spaces between option letters)
$UNits int — Define standard length
$Use macro [items] — Invoke macro if not empty where items

```
                             are its arguments
$Variate [length] id's
$VErify                      Switch to write each
                             line of executing
                             macro to current
                             output channel
$WArning                     Switch to print
                             warnings
$Weight [id]
$WHile scalar macro
                             Execute macro
                             repeatedly while
                             scalar is not 0
$Yvariate id                 Name independent
                             variable
```

APPENDIX II

DATA AND GLIM PROGRAMS FOR THE EXAMPLES

The data which appear at the beginning of each section below should be placed in a separate file to be read by the corresponding $DINput instruction.

The instructions, which always begin with the definition of a macro called TITL and end with $FINish, and which produce the output found in the corresponding chapter and section of the book, may either be typed in directly to GLIM or placed in a program file which is then read by the instruction $INput 5. $DINput 1 is used as the default data input channel and $OUtput 6 as the secondary output channel for text. If the user wishes to have the output directly on the screen, instead of written to a text file, the instructions $OUt 6 80 and $CAlculate %O=6 should be omitted. On the Commodore Amiga the primary input and output channels are both 9.

Chapter 1 One-Way Frequency Tables

1.1 A Time Trend Model

```
! STRESSFUL EVENTS - HABERMAN (1978, P.3)
!SUBJECTS REPORTING ONE EVENT
!1  2  3  4  5  6  7  8  9 10 11 12 13 14 15 16 17 18 MONTH
15 11 14 17  5 11 10  4  8 10  7  9 11  3  6  1  1  4

$Macro TITL !                                          define title
   STRESSFUL EVENTS - HABERMAN (1978, P.3)!
$Endmac!
$UNits 18!                                               read data
$DAta FREQ!
$PRint 'Load data'!
$DINput 1!
$Yvariate FREQ!                                       define model
$ERror P!
$PRint 'Load CHIT (TESTSTAT.glim)'!
$INput 12 CHIT!
$PRint 'Load RESP (GLIMPLOT.glim)'!
$INput 23 RESP!
$CAlculate MON=%GL(18,1)!              create required variables
```

```
: %O=6!                                  send macro output to text file
$OUt 6 80!                                     send output to text file
$PRint TITL :!
$Fit!                                            fit independence model
$Use CHIT %DV %DF!                                  calculate Chi-square
$CAlculate %D=%DV!        save values for difference in Chi-squares
: %E=%DF!
$Display ER!                                display and plot residuals
$PRint / TITL :!
$Use RESP!
$PRint / TITL :!
$Fit MON!                                 fit linear time trend model
$Use CHIT!                                         calculate Chi-squares
$CAlculate %DV=%D-%DV!
: %DF=%E-%DF!
$Use CHIT!
$Display ER!                                display and plot residuals
$PRint / TITL :!
$Use RESP!
$PRint / TITL :!
$PRint 'Observed and Fitted Values' :!                 plot regression
$Plot %FV FREQ MON!
$CAlculate F=%LOG(FREQ)!
: T=%LOG(%FV)!
$PRint : 'Linear Regression' :!
$Plot T F MON!
$FINish
```

1.3 A Symmetry Model

```
! SELF-CLASSIFICATION BY SOCIAL CLASS - HABERMAN (1978, P.24)
! LOWER WORKING MIDDLE UPPER CLASS
    72    714    655    41

$Macro TITL !                                             define title
   SELF-CLASSIFICATION BY SOCIAL CLASS - HABERMAN (1978, P.24)
$Endmac!
$Macro UCHI !                        macro for repeated instructions
   $Use CHIT!                                     calculate Chi-squares
   $CAlculate %DV=%D-%DV!
   : %DF=%E-%DF!
   $Use CHIT!
$$Endmac!
$UNits 4!                                                    read data
$DAta FREQ!
$PRint 'Load data'!
$DINput 1!
$Yvariate FREQ!                                           define model
$ERror P!
$PRint 'Load CHIT (TESTSTAT.glim)'!
$INput 12 CHIT!
$ASsign CLAS=1,-1,-1,1!                    create required variables
$CAlculate C1=2*(%GL(4,1)-2.5)!                          linear effect
: C2=(C1/2)**2-1.25!                                  quadratic effect
$OUt 6 80!                                     send output to text file
$PRint TITL :!
$Fit!                                            fit independence model
```

```
$Use CHIT %DV %DF!                           calculate Chi-square
$CAlculate %D=%DV!    save values for difference in Chi-squares
: %E=%DF!
$Display ER!          display parameter estimates and residuals
$Fit CLAS!                               fit pure quadratic model
$Use UCHI!
$Display ER!          display parameter estimates and residuals
$PRint / TITL!
$Fit C1+C2!                            fit linear+quadratic model
$Use UCHI!
$Display E!                        display parameter estimates
$FINish
```

1.4 Periodicity Models

```
! SUICIDES, USA, 1968 - HABERMAN (1978, P.51)
! JAN. FEB. MAR. APR. MAY  JUNE JULY AUG. SEPT OCT. NOV. DEC.
  1720 1712 1924 1882 1870 1680 1868 1801 1756 1760 1666 1733
31 29 31 30 31 30 31 31 30 31 30 31            ! DAYS IN THE MONTHS
1 1 2 2 2 3 3 3 4 4 4 1                        ! SEASONS

$Macro TITL !                                        define title
   SUICIDES, USA, 1968 - HABERMAN (1978, P.51)!
$Endmac!
$Macro UCHI !                     macro for repeated instructions
   $Use CHIT!                              calculate Chi-squares
   $CAlculate %DV=%D-%DV!
   : %DF=%E-%DF!
   $Use CHIT!
   $Display ER!                        display and plot residuals
   $PRint / TITL :!
   $Use RESP!
$$Endmac
$UNits 12!                                              read data
$DAta FREQ!                                    since data file has
$PRint 'Load data'!                              variables by line
$DINput 1!                                      instead of column,
$DAta DAYS!                                       must read each
$DINput 1!                                   variable separately
$DAta SEAS!
$DINput 1!
$Yvariate FREQ!                                      define model
$ERror P!
$Offset DAYS!                                       constant term
$FActor SEAS 4!
$PRint 'Load CHIT (TESTSTAT.glim)'!
$INput 12 CHIT!
$PRint 'Load TRAN (DESIGN.glim)'!
$INput 13 TRAN!
$PRint 'Load RESP (GLIMPLOT.glim)'!
$INput 23 RESP!
$Use TRAN SEAS S1 S2 S3!              create required variables
$CAlculate DAYS=%LOG(DAYS)!
: SIN=%SIN((2*%GL(%NU,1)-1)*%PI/12)!
: COS=%SQR(1-SIN*SIN)!
: %O=6!                          send macro output to text file
```

```
$OUt 6 80!                                   send output to text file
$PRint TITL :!
$Fit!                                        fit independence model
$Use CHIT %DV %DF!                               calculate Chi-square
$CAlculate %D=%DV!     save values for difference in Chi-squares
: %E=%DF!
$Display ER!                              display and plot residuals
$PRint / TITL :!
$Use RESP!
$PRint / TITL :!
$Fit SEAS!                fit seasons model using factor variable
$Display E!                             display parameter estimates
$Fit S1+S2+S3!   fit seasons model with conventional constraints
$Use UCHI!
$PRint / TITL :!
$Fit SIN+COS!                                fit sine-cosine model
$Use UCHI!
$PRint / TITL :!                                   plot regression
$PRint 'Observed and Fitted Values' :!
$CAlculate N=%GL(12,1)!
: T=%LOG(%FV)!
: F=%LOG(FREQ)!
$Plot %FV FREQ N!
$PRint 'Harmonic Model' :!
$Plot F T N!
$FINish
```

1.5 Local Effects

```
! SUICIDES (DURKHEIM) - HABERMAN (1978, P.87)
! MON. TUES WED. THUR FRI. SAT. SUN.
  1001 1035  982 1033  905  737  894

$Macro TITL !                                         define title
   SUICIDES (DURKHEIM) - HABERMAN (1978, P.87)!
$Endmac!
$UNits 7!                                                read data
$DAta FREQ!
$PRint 'Load data'!
$DINput 1!
$Yvariate FREQ!                                       define model
$ERror P!
$PRint 'Load CHIT (TESTSTAT.glim)'!
$INput 12 CHIT!
$PRint 'Load RESP (GLIMPLOT.glim)'!
$INput 23 RESP!
$CAlculate DAYS=2*(%GL(7,1)<=4)-1!     create required variables
: WEEK=(%GL(7,1)<=4)!
: %O=6!                           send macro output to text file
$OUt 6 80!                                 send output to text file
$PRint TITL :!
$Fit!                                      fit independence model
$Use CHIT %DV %DF!                             calculate Chi-square
$CAlculate %D=%DV!     save values for difference in Chi-squares
: %E=%DF!
$Display ER!                            display and plot residuals
$PRint / TITL :!
```

```
$Use RESP!
$PRint / TITL :!
$Fit DAYS!                         fit model with 2 periods within the week
$Use CHIT!                                            calculate Chi-squares
$CAlculate %DV=%D-%DV!
: %DF=%E-%DF!
$Use CHIT!
$Display ER!                                    display and plot residuals
$PRint / TITL :!
$Use RESP!
$PRint / TITL :!
$Weight WEEK!                           eliminate Friday and the weekend
$Fit!                                           refit independence model
$Use CHIT!                                           calculate Chi-square
$Display ER!                                display parameter estimates
$FINish
```

Chapter 2 Time and Causality

2.1 Retrospective Studies I

```
!BRITISH SOCIAL MOBILITY - GLASS (1954) - BISHOP ET AL (1975, P.100)
! 1   2   3   4   5   SON
 50  45   8  18   8 ! FATHER 1: 1 PROFESSIONAL & HIGH ADMINISTRATIVE
 28 174  84 154  55 ! FATHER 2: 2 MANAGERIAL, EXEC. & HIGH SUPERVIS.
 11  78 110 223  96 ! FATHER 3: 3 LOW INSPECTIONAL & SUPERVISORY
 14 150 185 714 447 ! FATHER 4: 4 ROUTINE NONMANUAL & SKILLED MANUAL
  3  42  72 320 411 ! FATHER 5: 5 SEMI- & UNSKILLED MANUAL

$Macro TITL !                                                define title
   BRITISH SOCIAL MOBILITY - GLASS (1954)!
$Endmac!
$UNits 25!                                                      read data
$DAta FREQ!
$PRint 'Load data'!
$DINput 1!
$Yvariate FREQ!                                              define model
$ERror P!
$FActor SON 5 FATH 5!
$PRint 'Load TRAN (DESIGN.glim)'!
$INput 13 TRAN IN44!
$PRint 'Load RESP (GLIMPLOT.glim)'!
$INput 23 RESP!
$CAlculate SON=%GL(5,1)!                        create required variables
: FATH=%GL(5,5)!
: %O=6!                                   send macro output to text file
$Use TRAN SON SON1 SON2 SON3 SON4!            conventional constraints
$Use TRAN FATH FAT1 FAT2 FAT3 FAT4!
$Use IN44 SON1 SON2 SON3 SON4 FAT1 FAT2 FAT3 FAT4!   interaction
$OUt 6 80!                                        send output to text file
$PRint TITL :!
$Fit SON+FATH!                                    fit independence model
$Display E!                                   display parameter estimates
$PRint / TITL :!                                refit independence model
$Fit SON1+SON2+SON3+SON4+FAT1+FAT2+FAT3+FAT4!
```

```
$Display ER!                              display and plot residuals
$PRint / TITL :!
$Use RESP!
$PRint / TITL :!
$Fit SON+FATH+SON.FATH!         fit interaction: saturated model
$Display E!                          display parameter estimates
$PRint / TITL :!
$Fit SON1+SON2+SON3+SON4+FAT1+FAT2+FAT3+FAT4+#I44!          refit
!                                              interaction model
$Display E!                          display parameter estimates
$FINish
```

2.2 Retrospective Studies II

```
! CLINIC USE (FIENBERG, 1977, P.92)
! YES NO  USE      ATTITUDE        VIRGIN
   23 23 !     ALWAYS WRONG         YES
   29 67 !    NOT ALWAYS WRONG      YES
  127 18 !     ALWAYS WRONG          NO
  112 15 !    NOT ALWAYS WRONG       NO


$Macro TITL !                                       define title
   CLINIC USE (FIENBERG, 1977, P.92)!
$Endmac!
$Macro UCHI !                    macro for repeated instructions
   $Use CHIT!                              calculate Chi-squares
   $CAlculate %DV=%D-%DV!
   : %DF=%E-%DF!
   $Use CHIT!
$$Endmac!
$UNits 8!                                              read data
$DAta FREQ!
$PRint 'Load data'!
$DINput 1!
$Yvariate FREQ!                                     define model
$ERror P!
$PRint 'Load CHIT (TESTSTAT.glim)'!
$INput 12 CHIT!
$CAlculate USE=3-%GL(2,1)*2!          create required variables
: ATTI=3-%GL(2,2)*2!               using conventional constraints
: VIRG=3-%GL(2,4)*2!
: UA=USE*ATTI!
: UV=USE*VIRG!
: AV=ATTI*VIRG!
$OUt 6 80!                              send output to text file
$PRint TITL :!
$Fit ATTI+VIRG+USE+AV!                    fit independence model
$Use CHIT %DV %DF!                          calculate Chi-square
$CAlculate %D=%DV!    save values for difference in Chi-squares
: %E=%DF!
$Display E!                          display parameter estimates
$Fit +UV!                                      fit virgin effect
$Use UCHI!
$Display E!                          display parameter estimates
$PRint / TITL :!
$Fit -UV+UA!                                 fit attitude effect
```

```
$Use UCHI!
$Display E!                                     display parameter estimates
$Fit +UV!                                        fit virgin+attitude effect
$Use UCHI!
$Display E!                                     display parameter estimates
$OUt!
$DElete FREQ ATTI VIRG USE!
$UNits 4!                    reread data after going to top of file
$DAta USE N!
$REWind 1!
$DINput 1!
$Yvariate USE!                                           redefine model
$ERror B N!
$CAlculate N=USE+N!                           create required variables
: ATTI=3-%GL(2,1)*2!
: VIRG=3-%GL(2,2)*2!
$OUt 6 80!                                     send output to text file
$PRint / TITL :!
$Fit!                                              fit independence model
$Use CHIT!
$CAlculate %D=%DV!      save values for difference in Chi-squares
: %E=%DF!
$Display E!                                     display parameter estimates
$Fit +VIRG!                                              fit virgin effect
$Use UCHI!
$Display E!                                     display parameter estimates
$PRint / TITL :!
$Fit -VIRG+ATTI!                                       fit attitude effect
$Use UCHI!
$Display E!                                     display parameter estimates
$Fit +VIRG!                                      fit virgin+attitude effect
$Use UCHI!
$Display E!                                     display parameter estimates
$FINish
```

2.3 Panel Studies

```
! MEMBERS OF THE LEADING CROWD  - BOYS (COLEMAN, 1964, P.171)
! FAV  UNF - ATTITUDE 1   MEMBER 1
 757  496 !                 YES
1071 1074 !                  NO
! YES   NO  - MEMBER 2   ATTITUDE 1 MEMBER 1
  598  159 !                FAV         YES
  353  143 !                UNF         YES
  259  812 !                FAV          NO
  182  892 !                UNF          NO
! FAV  UNF - ATTITUDE 2  ATTITUDE 1 MEMBER 1 MEMBER 2
  458  140 !                FAV         YES       YES
  171  182 !                UNF         YES       YES
  184   75 !                FAV          NO       YES
   85   97 !                UNF          NO       YES
  110   49 !                FAV         YES        NO
   56   87 !                UNF         YES        NO
  531  281 !                FAV          NO        NO
  338  554 !                UNF          NO        NO
```

```
$Macro TITL !                                             define title
   MEMBERS OF THE LEADING CROWD - BOYS (COLEMAN, 1964, P.171)!
$Endmac!
$Macro UCHI !                         macro for repeated instructions
   $Use CHIT!                                   calculate Chi-squares
   $CAlculate %DV=%D-%DV!
   : %DF=%E-%DF!
   $Use CHIT!
$$Endmac!
$UNits 2!                                                   read data
$DAta A1 A2!
$PRint 'Load data'!
$DINput 1!
$DAta 4 M1 M2!
$DINput 1!
$DAta 8 AA1 AA2!
$DINput 1!
$Yvariate A1!                                            define model
$ERror B N!
$PRint 'Load CHIT (TESTSTAT.glim)'!
$INput 12 CHIT!
$Variate 8 ATT1 MEM1 MEM2!
$CAlculate ATT1=3-%GL(2,1)*2!            create required variables
: MEM1=3-%GL(2,2)*2!
: MEM2=3-%GL(2,4)*2!
: MEM=3-%GL(2,1)*2!
: N=A1+A2!
: %O=6!                             send macro output to text file
$OUt 6 80!                                send output to text file
$PRint TITL :!
: 'Response Variable: ATT1' :!
$Fit!                                       fit independence model
$Use CHIT %DV %DF!                             calculate Chi-square
$Display E!                             display parameter estimates
$Fit +MEM!                                   fit membership effect
$Display E!                             display parameter estimates
$PRint / TITL :!
: 'Response Variable: MEM2' :!
$OUt!
$DElete MEM N!
$UNits 4!                                              redefine model
$ERror B N!
$Yvariate M1!
$CAlculate MEM=3-%GL(2,2)*2!             create required variables
: ATT=3-%GL(2,1)*2!
: N=M1+M2!
$OUt 6 80!                                send output to text file
$Fit!                                       fit independence model
$Use CHIT!                                     calculate Chi-square
$CAlculate %D=%DV!     save values for difference in Chi-squares
: %E=%DF!
$Display E!                             display parameter estimates
$Fit +MEM!                                   fit membership effect
$Use UCHI!
$Display E!                             display parameter estimates
$Fit +ATT!                                      add attitude effect
$Use UCHI!
$Display E!                             display parameter estimates
$PRint / TITL :!
```

```
: 'Response Variable: ATT2' :!
$OUt!
$DElete N!
$UNits 8!                                                  redefine model
$ERror B N!
$Yvariate AA1!
$OUt 9!
$PRint 'Load RESP (GLIMPLOT.glim)'!
$INput 23 RESP!
$CAlculate N=AA1+AA2!                          create required variables
$OUt 6 80!                                      send output to text file
$Fit!                                            fit independence model
$Use CHIT!                                           calculate Chi-square
$CAlculate %D=%DV!     save values for difference in Chi-squares
: %E=%DF!
$Display ER!                                display and plot residuals
$PRint / TITL :!
$Use RESP!
$PRint / TITL :!
$Fit +ATT1+MEM1+MEM2!                     fit attitude + 2 memberships
$Use UCHI!
$Display ER!                                display and plot residuals
$PRint / TITL :!
$Use RESP!
$PRint / TITL :!
$Fit -MEM1!                             remove membership 1 from fit
$Use UCHI!
$Display ER!                                display and plot residuals
$PRint / TITL :!
$Use RESP!
$FINish
```

2.4 First Order Markov Chains

```
! ONE STEP TRANSITIONS - VOTERS IN ERIE COUNTY, 1940 (GOODMAN,
! 1962)
! R   D   U
!   JUNE
 125   5  16 !  R                            PARTIES:   D - DEMOCRAT
   7 106  15 !  D  MAY                                  R - REPUBLICAN
  11  18 142 !  U                                       U - UNDECIDED
!   JULY
 124   3  16 !  R
   6 109  14 !  D  JUNE
  22   9 142 !  U
!  AUGUST
 146   2   4 !  R
   6 111   4 !  D  JULY
  40  36  96 !  U
!  SEPTEMBER
 184   1   7 !  R
   4 140   5 !  D  AUGUST
  10  12  82 !  U
!   OCTOBER
 192   1   5 !  R
   2 146   5 !  D  SEPTEMBER
  11  12  71 !  U
```

```
$Macro TITL !                                                  define title
   ONE STEP TRANSITIONS - VOTERS IN ERIE COUNTY, 1940 (GOODMAN,!
   1962)!
$Endmac!
$UNits 45!                                                        read data
$DAta FREQ!
$PRint 'Load data'!
$DINput 1!
$Yvariate FREQ!                                                define model
$ERror P!
$FActor T1 3 T2 3 TIME 5!
$PRint 'Load CHIT (TESTSTAT.glim)'!
$INput 12 CHIT!
$PRint 'Load MPCT (CONTTAB.glim)'!
$INput 15 MPCT!
$PRint 'Load RESP (GLIMPLOT.glim)'!
$INput 23 RESP!
$CAlculate T1=%GL(3,3)!                           create required variables
: T2=%GL(3,1)!
: TIME=%GL(5,9)!
: %O=6!                                      send macro output to text file
$OUt 6 80!                                         send output to text file
$PRint / TITL :!
$Use MPCT T1 T2 TIME!                                  test for stationarity
$Use CHIT %DV %DF!                                     calculate Chi-square
$Display E!                                    display and plot residuals
$PRint / TITL :!
$Display R!
$PRint / TITL :!
$Use RESP!
$OUt!                                                          stop output
$DElete PW %RE T1 T2 TIME!
$UNits 18!                                                 redefine model
$ERror P!
$Yvariate F!
$FActor T1 3 T2 3 TIME 2!
$Weight PW!
$Variate 45 K!                               define variable with old size
$CAlculate K=%GL(45,1)!                           create required variables
: F(K*(K<=18))=FREQ!                                        first 3 months
: T1=%GL(3,3)!
: T2=%GL(3,1)!
: TIME=%GL(2,9)!
: PW=1!
$OUt 6 80!                                         send output to text file
$PRint / TITL :!
: 'May-June-July Period' : :!
$Use MPCT T1 T2 TIME!           test stationarity of first 3 months
$Use CHIT!                                             calculate Chi-square
$Display E!                                    display and plot residuals
$PRint / TITL :!
$Display R!
$PRint / TITL :!
$Use RESP!
$OUt!
$CAlculate F((K-27)*(K>=28))=FREQ!create variable: last 3 months
$OUt 6 80!
$PRint / TITL :!
```

```
: 'August-September-October Period' : :!
$Use MPCT!              test for stationarity of last 3 months
$Use CHIT!                                   calculate Chi-square
$Display E!                           display and plot residuals
$PRint / TITL :!
$Display R!
$PRint / TITL :!
$Use RESP!
$FINish
```

2.5 Second Order Markov Chains

```
! TWO STEP TRANSITIONS - VOTERS IN ERIE COUNTY, 1940 (GOODMAN,
! 1962)
!  TIME T                                 PARTIES: D - DEMOCRAT
! R   D   U   TIME T-2  TIME T-1                   R - REPUBLICAN
 557   6  16 !   R                                 U - UNDECIDED
  18   0   5 !   D        R
  71   1  11 !   U
   3   8   0 !   R
   9 435  22 !   D        D
   6  63   6 !   U
  17   5  21 !   R
   4  10  24 !   D        U
  62  54 346 !   U

$Macro TITL !                                         define title
   TWO STEP TRANSITIONS - VOTERS IN ERIE COUNTY, 1940 (GOODMAN,!
   1962)!
$Endmac
$UNits 27!                                               read data
$DAta FREQ!
$PRint 'Load data'!
$DINput 1!
$Yvariate FREQ!                                       define model
$ERror P!
$FActor T1 3 T2 3 T3 3!
$PRint 'Load CHIT (TESTSTAT.glim)'!
$INput 12 CHIT!
$PRint 'Load RESP (GLIMPLOT.glim)'!
$INput 23 RESP!
$CAlculate T1=%GL(3,3)!               create required variables
: T2=%GL(3,9)!
: T3=%GL(3,1)!
: %O=6!                           send macro output to text file
$OUt 6 80!                              send output to text file
$PRint TITL :!
$Fit T1*T2+T2*T3!                      test if first order process
$Use CHIT %DV %DF!                           calculate Chi-square
$Display ER!                          display and plot residuals
$PRint / TITL :!
$Use RESP!
$FINish
```

Chapter 3 Metric Variables

3.1 Time Trends

```
! ATTITUDE TO CRIMINALS 1972-1975 - HABERMAN (1978, P.120)
! 1972 1973 1974 1975     ATTITUDE
   105   68   42   61 !  TOO HARSHLY
  1066 1092  580 1174 !  NOT HARSHLY ENOUGH
   265  196   72  144 !  ABOUT RIGHT
   173  138   51  104 !  DON'T KNOW
     4   10    8    7 !  NO ANSWER

$Macro TITL !                                      define title
   ATTITUDE TO CRIMINALS 1972-1975 - HABERMAN (1978, P.120)!
$Endmac!
$Macro CH1 !                       macro for repeated instructions
   $Use CHIT                                 calculate Chi-squares
   $CAlculate %D=%DV!
   : %E=%DF!
   $Display ER!                         display and plot residuals
   $PRint / TITL :!
   $Use RESP!
$$Endmac!
$Macro CH2 !                       macro for repeated instructions
   $Use CHIT!                                calculate Chi-squares
   $CAlculate %DV=%D-%DV!
   : %DF=%E-%DF!
   $Use CHIT!
   $Display ER!                         display and plot residuals
   $PRint / TITL :!
   $Use RESP!
$$Endmac!
$UNits 20!                                               read data
$DAta FREQ!
$PRint 'Load data'!
$DINput 1!
$Yvariate FREQ!                                       define model
$ERror P!
$FActor YEAR 4 ATTI 5!
$PRint 'Load CHIT (TESTSTAT.glim)'!
$INput 12 CHIT!
$PRint 'Load ORTH (DESIGN.glim)'!
$INput 13 ORTH!
$PRint 'Load RESP (GLIMPLOT.glim)'!
$INput 23 RESP!
$CAlculate YEAR=%GL(4,1)!             create required variables
: ATTI=%GL(5,4)!
: PW=1!
: %O=6!                           send macro output to text file
$Use ORTH YEAR YRL YRQ YRC!
$Argument CHIT %DV %DF!
$OUt 6 80!                               send output to text file
$PRint TITL :!
$Fit YEAR+ATTI!                          fit independence model
$Use CH1!
$PRint / TITL :!
```

```
$Fit +ATTI.YRL!                              fit linear effect of year
$Use CH2!
$CAlculate W=(ATTI/=2)!eliminate attitude 2 (not harshly enough)
$Weight W!
$PRint / TITL :!
$Fit -ATTI.YRL!                               refit independence model
$Use CH1!
$CAlculate W=W*(ATTI/=5)!        eliminate attitude 5 (no answer)
$PRint / TITL :!
$F.!                                          refit independence model
$Use CH2!
$FINish
```

3.2 Model Simplification

```
! ATTITUDE TO WOMEN STAYING AT HOME - HABERMAN (1979, P.312)
!AGREE DISAGREE ATTITUDE  SEX EDUC
    4       2 !                M   0
    4       2 !                F   0
    2       0 !                M   1
    1       0 !                F   1
    4       0 !                M   2
    0       0 !                F   2
    6       3 !                M   3
    6       1 !                F   3
    5       5 !                M   4
   10       0 !                F   4
   13       7 !                M   5
   14       7 !                F   5
   25       9 !                M   6
   17       5 !                F   6
   27      15 !                M   7
   26      16 !                F   7
   75      49 !                M   8
   91      36 !                F   8
   29      29 !                M   9
   30      35 !                F   9
   32      45 !                M  10
   55      67 !                F  10
   36      59 !                M  11
   50      62 !                F  11
  115     245 !                M  12
  190     403 !                F  12
   31      70 !                M  13
   17      92 !                F  13
   28      79 !                M  14
   18      81 !                F  14
    9      23 !                M  15
    7      34 !                F  15
   15     110 !                M  16
   13     115 !                F  16
    3      29 !                M  17
    3      28 !                F  17
    1      28 !                M  18
    0      21 !                F  18
    2      13 !                M  19
    1       2 !                F  19
```

```
   3     20 !                   M  20
   2      4 !                   F  20

$Macro TITL !                                          define title
   ATTITUDE TO WOMEN STAYING AT HOME - HABERMAN (1979, P.312)!
$Endmac!
$Macro UCHI !                    macro for repeated instructions
   $Use CHIT!                                calculate Chi-squares
   $CAlculate %DV=%D-%DV!
   : %DF=%E-%DF!
   $Use CHIT!
   $CAlculate %D=%D-%DV!
   : %E=%E-%DF!
   $Display E!                          display and plot residuals
   $PRint / TITL :!
   $Display R!
   $PRint / TITL :!
   $Use RESP!
$$Endmac
$UNits 42!                                               read data
$DAta A D!
$PRint 'Load data'!
$DINput 1!
$Yvariate A!                                          define model
$ERror B N!
$FActor EDUC 21!
$PRint 'Load CHIT (TESTSTAT.glim)'!
$INput 12 CHIT!
$PRint 'Load ORTH (DESIGN.glim)'!
$INput 13 ORTH!
$PRint 'Load RESP (GLIMPLOT.glim)'!
$INput 23 RESP!
$Weight PW!
$CAlculate SEX=3-2*%GL(2,1)!          create required variables
: EDUC=%GL(21,2)!
: N=A+D!
: PW=1!
: PW(6)=0!               eliminate category with no observations
: N(6)=1!
: %O=6!                          send macro output to text file
$Use ORTH EDUC EDL EDQ EDC!    calculate orthogonal polynomials
$CAlculate ESL=SEX*EDL!                               interactions
: ESQ=SEX*EDQ!
$OUt 6 80!                             send output to text file
$PRint TITL :!
$Fit!                                     fit independence model
$Use CHIT %DV %DF!                          calculate Chi-square
$CAlculate %D=%DV!     save values for difference in Chi-squares
: %E=%DF!
$Display ER!                            display and plot residuals
$PRint / TITL :!
$Use RESP!
$PRint / TITL :!
$Fit +SEX!                                          fit sex effect
$Use UCHI!
$PRint / TITL :!
$Fit -SEX+EDUC!                      fit complete education effect
$Use UCHI!
$PRint / TITL :!
```

```
$Fit -EDUC+EDL!                              fit linear education effect
$Use CHIT!                                           calculate Chi-square
$CAlculate %D=%DV!      save values for difference in Chi-squares
: %E=%DF!
$Display E!                                  display and plot residuals
$PRint / TITL :!
$Display R!
$PRint / TITL :!
$Use RESP!
$PRint / TITL :!
$Fit +SEX+ESL!       add sex + sex x linear education interaction
$Use UCHI!
$CAlculate PW=(EDUC>=7)!     eliminate lower levels of education
$Weight PW!
$PRint / TITL :!
$Fit SEX+EDL+ESL!     refit sex + linear education + interaction
$Use CHIT!
$Display E!                                  display and plot residuals
$PRint / TITL :!
$Display R!
$PRint / TITL :!
$Use RESP!
$PRint / TITL :!                                        plot regression
: 'Observed and Fitted Values' :!
$CAlculate F=%FV/%BD!
: O=A/%BD!
: EDUC=EDUC-1!                         reset education to initial values
: SEX=%GL(2,1)!        setup for different characters for 2 sexes
$FActor SEX 2!                                                  in plot
$Plot O F EDUC 'MF mf' SEX!
$FINish
```

Chapter 4 Ordinal variables

4.1 The Log Multiplicative Model I

```
! CRIMINAL CASES IN N. CAROLINA, OFFENCE, COUNTY, RACE (UPTON,
! 1978, P.104
!     OUTCOME OF CASE
! NO PROS. GUILTY  NOT GUIL.  OFFENCE     RACE    COUNTY
     33       8         4 !    DRINKING    BLACK   DURHAM
     10      10         3 !    VIOLENCE    BLACK   DURHAM
      9       8         2 !    PROPERTY    BLACK   DURHAM
      4       2         1 !    TRAFFIC     BLACK   DURHAM
     32       3         0 !    SPEEDING    BLACK   DURHAM
      5      10         1 !    DRINKING    BLACK   ORANGE
      5       5         5 !    VIOLENCE    BLACK   ORANGE
     11       5         3 !    PROPERTY    BLACK   ORANGE
     12       6         1 !    TRAFFIC     BLACK   ORANGE
     20       3         2 !    SPEEDING    BLACK   ORANGE
     53       2         2 !    DRINKING    WHITE   DURHAM
      7       8         1 !    VIOLENCE    WHITE   DURHAM
     10       5         2 !    PROPERTY    WHITE   DURHAM
     16       3         2 !    TRAFFIC     WHITE   DURHAM
     87       5         3 !    SPEEDING    WHITE   DURHAM
```

```
      14        2        0 !     DRINKING   WHITE  ORANGE
       1        5        7 !     VIOLENCE   WHITE  ORANGE
       5        4        0 !     PROPERTY   WHITE  ORANGE
      13       13        1 !     TRAFFIC    WHITE  ORANGE
      98       16        7 !     SPEEDING   WHITE  ORANGE

$Macro PRES !                        macro for repeated instructions
   $Display E!                            display and plot residuals
   $PRint / TITL :!
   $Display R!
   $PRint / TITL :!
   $Use RESP!
$$Endmac!
$UNits 60!                                                 read data
$DAta FREQ!
$PRint 'Load data'!
$DINput 1!
$Yvariate FREQ!                                         define model
$ERror P!
$FActor OUT 3 IND 20 OFF 5 COUN 2 RACE 2!
$PRint 'Load L1OV (ORDVAR.glim)'!
$INput 16 L1OV!
$Macro TITL !                                           define title
   CRIMINAL CASES IN N. CAROLINA, OFFENCE, COUNTY, RACE!
   (UPTON, 1978, P.104)!
$Endmac!
$CAlculate OUT=%GL(3,1)!                create required variables
: OFF=%GL(5,3)!
: COUN=%GL(2,15)!
: RACE=%GL(2,30)!
: IND=%GL(20,3)!
: %O=6!                               send macro output to text file
: %R=1!                           display and plot residuals in macros
$Use L1OV IND OUT!                        fit log multiplicative model
$PRint / TITL :!
$Fit OUT+IND+ZZ1_.OFF+ZZ1_.COUN+ZZ1_.RACE!              refit without
!                                                        interactions
$Use CHIT %DV %DF!                               calculate Chi-square
$Use PRES!
$PRint / TITL :!
$Fit +ZZ1_.OFF.RACE!                  add offence x race interaction
$Use CHIT!                                       calculate Chi-square
$Use PRES!
$FINish
```

4.2 The Log Multiplicative Model II

```
! SCHIZOPHRENIC PATIENTS IN LONDON (FIENBERG, 1977, P.55)
!    1    2    3 - VISITS LENGTH (YEARS)       VISITS:
   43    6    9 !           2-10      GOES HOME OR VISITED REGULARLY
   16   11   18 !          10-20      VISITED < ONCE A MONTH & DOES
!                                       NOT GO HOME
    3   10   16 !           >20       NEVER VISITED & NEVER GOES HOME

$UNits 9!                                                  read data
$DAta FREQ!
$PRint 'Load data'!
```

```
$DINput 1!
$Yvariate FREQ!                                         define model
$ERror P!
$FActor VIS 3 LENG 3!
$PRint 'Load L2OV (ORDVAR.glim)'!
$INput 16 L2OV!
$Macro TITL !                                           define title
  SCHIZOPHRENIC PATIENTS IN LONDON (FIENBERG, 1977, P.55)!
$Endmac!
$CAlculate VIS=%GL(3,1)!
: LENG=%GL(3,3)!
: %O=6!                              send macro output to text file
: %R=1!                        display and plot residuals in macros
$Use L2OV VIS LENG!                     fit log multiplicative model
$FINish
```

4.3 The Proportional Odds Model

```
$UNits 9!                                                  read data
$DAta FREQ!
$PRint 'Load data'!
$DINput 1!
$Yvariate FREQ!
$ERror P!
$FActor VIS 3!
$PRint 'Load ORTH (DESIGN.glim)'!
$INput 13 ORTH!
$PRint 'Load POOV (ORDVAR.glim)'!
$INput 16 POOV!
$Macro TITL !                                           define title
  SCHIZOPHRENIC PATIENTS IN LONDON (FIENBERG, 1977, P.55)!
$Endmac!
$CAlculate VIS=%GL(3,1)!                 create required variables
: LENG=%GL(3,3)!
: %N=1!
: %K=3!
: %L=3!
: %O=6!                              send macro output to text file
$Use ORTH LENG LENL LENQ!
$OUt 6 80!                                 send output to text file
$PRint TITL :!
$Use POOV FREQ LENL!                   fit proportional odds model
$OUt 9!
$PRint 'Load RESP (GLIMPLOT.glim)'!
$INput 23 RESP!      load macro now, since model changed by POOV
$OUt 6 80!
$PRint / TITL :!                                     plot residuals
$Use RESP!
$FINish
```

4.4 The Continuation Ratio Model

```
$UNits 9!                                                  read data
$DAta FREQ!
```

```
$PRint 'Load data'!
$DINput 1!
$PRint 'Load ORTH (DESIGN.glim)'!
$INput 13 ORTH!
$PRint 'Load L1OV (ORDVAR.glim)'!
$INput 16 CROV!
$PRint 'Load RESP (GLIMPLOT.glim)'!
$INput 23 RESP!
$Macro TITL !                                           define title
   SCHIZOPHRENIC PATIENTS IN LONDON (FIENBERG, 1977, P.55)!
$Endmac!
$CAlculate VIS=%GL(3,1)!                create required variables
: LENG=%GL(3,3)!
: %L=3!
: %K=3!
: %O=6!                              send macro output to text file
$OUt 6 80!                                send output to text file
$PRint TITL :!
$Use CROV!                              fit continuation ratio model
$Display R!                               display and plot residuals
$PRint / TITL :!
$Use RESP!
$Use ORTH ZZ2_ LENL LENQ!         calculate orthogonal polynomial
$PRint / TITL :!
$Fit ZZ1_+LENL!              fit model with linear length of visit
$Use CHIT!                                     calculate Chi-square
$Display ER!                              display and plot residuals
$PRint / TITL :!
$Use RESP!
$FINish
```

Chapter 5 Zero Frequencies and Incomplete Tables

5.1 Sampling Zeroes

```
! SWEDISH ELECTIONS 1964 AND 1970 (FINGLETON, 1984, P.138)
! COMM  SD    C    P   CON - 1970   PARTIES: COMM - COMMUNIST
   22   27    4    1    0 ! COMM          SD   - SOCIAL DEMOCRAT
   16  861   57   30    8 ! SD            C    - CENTRE
    4   26  248   14    7 ! C   1964      P    - PEOPLE'S
    8   20   61  201   11 ! P             CON  - CONSERVATIVE
    0    4   31   32  140 ! CON

$Macro TITL !                                           define title
   SWEDISH ELECTIONS 1964 AND 1970 (FINGLETON, 1984, P.138)!
$Endmac!
$Macro DIER !                         macro for repeated instructions
   $Display E!        display parameter estimates and residuals
   $PRint / TITL :!
   $Display R!
$$Endmac!
$UNits 25!                                                 read data
$DAta FREQ!
$PRint 'Load data'!
$DINput 1!
```

```
$Yvariate FREQ!                                         define model
$ERror P!
$FActor V70 5 V64 5!
$PRint 'Load DFCT (CONTTAB.glim)'!
$INput 15 DFCT!
$CAlculate V70=%GL(5,1)!                   create required variables
: V64=%GL(5,5)!
: PW=1!
: V704=V64*V70!
$OUt 6 80!                                 send output to text file
$PRint TITL :!
$Fit V70+V64+V70.V64!                           fit saturated model
$Use DIER!
$PRint / TITL :!
$Use DFCT!                                              correct d.f.
$Use DIER!
$Weight PW!
$PRint / TITL :!
$Fit V70+V64+V704!                    fit linear interaction model
$Use DIER!
$PRint / TITL :!
$Use DFCT!                                              correct d.f.
$Use DIER!
$FINish
```

5.2 Incomplete Tables and Quasi-Independence

```
! HEALTH PROBLEMS (FIENBERG, 1977, P.116)
!   PROBLEM SEX AGE
 4      1     1   1 !     PROBLEMS: 1 - SEX & REPRODUCTION
42      3     1   1 !               2 - MENSTRUATION
57      4     1   1 !               3 - HOW HEALTHY I AM
 2      1     1   2 !               4 - NOTHING
 7      3     1   2 !          SEX: 1 - MALE
20      4     1   2 !               2 - FEMALE
 9      1     2   1 !          AGE: 1 - 12-15
 4      2     2   1 !               2 - 16-17
19      3     2   1 !
71      4     2   1 !
 7      1     2   2 !
 8      2     2   2 !
10      3     2   2 !
31      4     2   2 !

$Macro TITL !                                           define title
  HEALTH PROBLEMS (FIENBERG, 1977, P.116)!
$Endmac!
$Macro UCHI !                      macro for repeated instructions
   $Use CHIT!                                calculate Chi-squares
   $CAlculate %DV=%D-%DV!
   : %DF=%E-%DF!
   $Use CHIT!
$$Endmac!
$UNits 14!                                                 read data
$DAta FREQ PROB SEX AGE!
$PRint 'Load data'!
$DINput 1!
```

```
$Yvariate FREQ!                                              define model
$ERror P!
$PRint 'Load CHIT (TESTSTAT.glim)'!
$INput 12 CHIT!
$PRint 'Load TRAN (DESIGN.glim)'!
$INput 13 TRAN!
$PRint 'Load RESP (GLIMPLOT.glim)'!
$INput 23 RESP!
$Use TRAN PROB PRO1 PRO2 PRO3!
$CAlculate SEX=3-2*SEX!                          create required variables
: AGE=3-2*AGE!                               using conventional constraints
: SA=SEX*AGE!                                                 interactions
: AP1=AGE*PRO1!
: AP2=AGE*PRO2!
: AP3=AGE*PRO3!
: SP1=SEX*PRO1!
: SP3=SEX*PRO3!
: %O=6!                                      send macro output to text file
$OUt 6 80!                                         send output to text file
$PRint TITL :!
$Fit PRO1+PRO2+PRO3+SEX+AGE+SA!                      fit independence model
$Use CHIT %DV %DF!                                    calculate Chi-square
$CAlculate %D=%DV!      save values for difference in Chi-squares
: %E=%DF!
$Display ER!                                  display and plot residuals
$PRint / TITL :!
$Use RESP!
$PRint / TITL :!
$Fit +SP1+SP3!                                              fit sex effect
$Use UCHI!
$Display ER!                                  display and plot residuals
$PRint / TITL :!
$Use RESP!
$PRint / TITL :!
$Fit -SP1-SP3+AP1+AP2+AP3!                                  fit age effect
$Use UCHI!
$CAlculate %D=%D-%DV!
: %E=%E-%DF!
$Display ER!                                  display and plot residuals
$PRint / TITL :!
$Use RESP!
$FINish
```

5.3 Population Estimation

```
! FORMAL VOLUNTEER ORGANIZATIONS (BISHOP ET AL, 1975, P.243)
!NO.  CENSUS NEWSPAPER TELEPHONE
  4      1       1          1 !
  1      2       1          1 !      1 - YES
  8      1       2          1 !      2 -  NO
  2      2       2          1 !
 16      1       1          2 !
 49      2       1          2 !
113      1       2          2 !
  0      2       2          2 !
```

```
$Macro TITL !                                              define title
   FORMAL VOLUNTEER ORGANIZATIONS (BISHOP ET AL, 1975, P.243)!
$Endmac!
$Macro EST !                          macro for repeated instructions
   $CAlculate %F=%CU(%FV)!            calculate population estimate
   : %V=%SQR(%FV(8)*%F/(%FV(1)+%FV(2)+%FV(3)+%FV(5)))!
   $PRint 'Estimated total =' *-4 %F ' with s.d. =' %V :!
$$Endmac!
$UNits 8!                                                     read data
$DAta FREQ CENS NEWS TELE!
$PRint 'Load data'!
$DINput 1!
$Yvariate FREQ!                                            define model
$ERror P!
$Weight PW!
$FActor CENS 2 NEWS 2 TELE 2!
$PRint 'Load CHIT (TESTSTAT.glim)'!
$INput 12 CHIT!
$CAlculate PW=1!                              create required variables
: PW(8)=0!                                eliminate impossible category
$OUt 6 80!                                    send output to text file
$PRint TITL :!
$Fit CENS*NEWS*TELE-CENS.NEWS.TELE-CENS.NEWS!            fit full model
!          except for newspaper x census and 3-way interactions
$Use CHIT %DV %DF!                                calculate Chi-square
$Display E!                               display parameter estimates
$Use EST!
$Fit -NEWS.TELE!         remove newspaper x telephone interaction
$Use CHIT!
$Display E!                               display parameter estimates
$Use EST!
$FINish
```

5.4 Social Mobility

```
! MIGRANT BEHAVIOUR - FINGLETON (1984, P.142)
! 1     2    3    4  REGION - 1971
 118   12    7   23 !  1         1 - CENTRAL CLYDESDALE
  14 2127   86  130 !  2 - 1966 2 - URBAN LANCASHIRE AND YORKSHIRE
   8   69 2548  107 !  3         3 - WEST MIDLANDS
  12  110   88 7712 !  4         4 - GREATER LONDON

$UNits 16!                                                    read data
$DAta FREQ!
$PRint 'Load data'!
$DINput 1!
$Yvariate FREQ!                                            define model
$ERror P!
$FActor M66 4  M71 4!
$PRint 'Load SMCT (CONTTAB.glim)'!
$INput 15 SMCT!
$Macro TITL !                                              define title
   MIGRANT BEHAVIOUR - FINGLETON (1984, P.142)!
$Endmac!
$CAlculate M71=%GL(4,1)!                      create required variables
: M66=%GL(4,4)!
: %R=1!                          display and plot residuals in macros
```

```
: %O=6!                                  send macro output to text file
$OUt 6 80!                                    send output to text file
$Use SMCT M66 M71!                                 fit mobility models
$FINish
```

5.5 The Bradley-Terry Model

```
! PREFERENCE FOR COLLECTIVE FACILITIES IN DENMARK (ANDERSEN, 1980,
! P.357)
!  1  2  3  4  5  6     NOT PREFERRED
   0 29 25 22 17  9 ! 1
  49  0 35 34 16 14 ! 2
  50 42  0 40 22 15 ! 3  PREFERRED
  54 43 37  0 33 16 ! 4
  61 61 54 44  0 27 ! 5
  69 64 63 62 51  0 ! 6

$Macro TITL !                                             define title
   PREFERENCE FOR COLLECTIVE FACILITIES IN DENMARK (ANDERSEN,!
   1980, P.357)!
$Endmac!
$UNits 36!                                                   read data
$DAta FREQ!
$PRint 'Load data'!
$DINput 1!
$Yvariate FREQ!                                           define model
$ERror P!
$FActor NOT 6 PREF 6!
$PRint 'Load BTCT (CONTTAB.glim)'!
$INput 15 BTCT!
$PRint 'Load RESP (GLIMPLOT.glim)'!
$INput 23 RESP!
$CAlculate PREF=%GL(6,6)!                  create required variables
: NOT=%GL(6,1)!
: %O=6!                                send macro output to text file
$OUt 6 80!                                  send output to text file
$PRint TITL :!
$Use BTCT NOT PREF!                          fit Bradley-Terry model
$PRint / TITL :!                          display and plot residuals
$Display R!
$PRint / TITL :!
$Use RESP!
$FINish
```

5.6 Guttman Scales

```
! ROLE CONFLICT (FIENBERG, 1977, P.126)
!
 42 !  1  1  1  1
 23 !  1  1  1  2
  6 !  1  1  2  1
 25 !  1  1  2  2
  6 !  1  2  1  1
 24 !  1  2  1  2
```

```
  7 !  1  2  2  1
 38 !  1  2  2  2
  1 !  2  1  1  1
  4 !  2  1  1  2
  1 !  2  1  2  1
  6 !  2  1  2  2
  2 !  2  2  1  1
  9 !  2  2  1  2
  2 !  2  2  2  1
 20 !  2  2  2  2

$Macro TITL !                                        define title
  ROLE CONFLICT (FIENBERG, 1977, P.126)!
$Endmac!
$UNits 16!                                              read data
$DAta FREQ!
$PRint 'Load data'!
$DINput 1!
$Yvariate FREQ!                                      define model
$ERror P!
$PRint 'Load CHIT (TESTSTAT.glim)'!
$INput 12 CHIT!
$PRint 'Load RESP (GLIMPLOT.glim)'!
$INput 23 RESP!
$Weight PW!
$CAlculate Q1=2*%GL(2,8)-3!           create required variables
: Q2=2*%GL(2,4)-3!
: Q3=2*%GL(2,2)-3!
: Q4=2*%GL(2,1)-3!
: PW=1!
: PW(1)=0!                         eliminate individuals on scale
: PW(2)=0!
: PW(4)=0!
: PW(8)=0!
: PW(16)=0!
: %O=6!                            send macro output to text file
$OUt 6 80!                               send output to text file
$PRint TITL :!
$Fit Q1+Q2+Q3+Q4!                          fit independence model
$Use CHIT %DV %DF!                           calculate Chi-square
$Display E!
$EXTract %PE!
$CAlculate P=1/(1+%EXP(2*%PE))!
: %A=P(2)!            probability of chance yes to each question
: %B=P(3)!
: %C=P(4)!
: %D=P(5)!
: %N=%CU(FREQ)!
: %E=%FV(3)/%N/%A/%B/(1-%C)/%D!     probability of chance answer
: %F=(%YV(1)-%FV(1))/%N!                 probabilities on scale
: %G=(%YV(2)-%FV(2))/%N!
: %H=(%YV(4)-%FV(4))/%N!
: %I=(%YV(8)-%FV(8))/%N!
: %J=(%YV(16)-%FV(16))/%N!
$PRint : 'Probability of replying by chance is' %E :!
: 'Probabilities of replying yes to each question by chance are'
: %A %B %C %D:!
: 'Probabilities of replying on the Guttman scale are'!
: %F %G %H %I %J!
```

```
$PRint / TITL :!                              display and plot residuals
$Display R!
$PRint / TITL :!
$Use RESP!
$FINish
```

Chapter 6 Patterns

6.1 Extremity Models

```
! OXFORD SHOPPING SURVEY (FINGLETON, 1984, P.10)
!DISAGREE  TEND TO     IN      TEND TO  AGREE - GROCERY SHOPPING
!          DISAGREE  BETWEEN    AGREE                  IS TIRING
    55        11        16        17       100  ! NO CAR AVAILABLE
   101         7        18        23       103  ! SOMETIMES CAR
!                                                       AVAILABLE
    91        20        25        16        77  ! CAR ALWAYS
!                                                       AVAILABLE

$Macro TITL !                                           define title
   OXFORD SHOPPING SURVEY (FINGLETON, 1984, P.10)!
$Endmac
$UNits 15!                                                 read data
$DAta FREQ!
$PRint 'Load data'!
$DINput 1!
$Yvariate FREQ!                                         define model
$ERror P!
$FActor TIRE 5 CAR 3 EX4 3 EX2 2!
$PRint 'Load CHIT (TESTSTAT.glim)'!
$INput 12 CHIT!
$PRint 'Load RESP (GLIMPLOT.glim)'!
$INput 23 RESP!
$CAlculate TIRE=%GL(5,1)!                create required variables
: CAR=%GL(3,5)!
: EX2=1!
: EX2(1)=EX2(15)=2!                              2 diagonal corners
: EX4=EX2!
: EX4(5)=EX4(11)=3!                              opposite 2 corners
: PW=1!
: %O=6!                            send macro output to text file
$OUt 6 80!                               send output to text file
$PRint TITL :!
: 'Independence Model' :!
$Fit TIRE+CAR!                            fit independence model
$Use CHIT %DV %DF!                          calculate Chi-square
$Display ER!                              display and plot residuals
$PRint / TITL :!
$Use RESP!
$PRint / TITL :!
: 'Extreme Ends Model' :!
$Fit +EX2!
$Use CHIT!                                  calculate Chi-square
$Display ER!                              display and plot residuals
$PRint / TITL :!
```

```
$Use RESP!
$PRint / TITL :!
: 'Four Corners Model' :!
$Fit -EX2+EX4!
$Use CHIT!                                         calculate Chi-square
$Display ER!                                  display and plot residuals
$PRint / TITL :!
$Use RESP!
$FINish
```

6.2 Symmetry Models

```
! BELGIAN ELECTIONS - 1981-1985 - VOTING CHANGES
! PS PRL PSC ECO PCB  BN          PARTIES:
 281  14   9  16   4   4 ! PS     SOCIALIST
  12 164  13   4   1   6 ! PRL    LIBERAL
   5  10 121   8   1   1 ! PSC    SOCIAL-CHRISTIAN
   6   0   1  50   0   1 ! ECO    ECOLOGY
   1   0   0   2  14   0 ! PCB    COMMUNIST
   2   1   0   0   0  11 ! BN     BLANK BALLOT

$UNits 36!                                                     read data
$DAta FREQ!
$PRint 'Load data'!
$DINput 1!
$Yvariate FREQ!                                             define model
$ERror P!
$FActor V81 6 V85 6!
$PRint 'Load SYCT (CONTTAB.glim)'!
$INput 15 SYCT MHCT!
$Macro TITL !                                               define title
   BELGIAN ELECTIONS - 1981-1985 - VOTING CHANGES!
$Endmac
$CAlculate V81=%GL(6,1)!                       create required variables
: V85=%GL(6,6)!
: %O=6!                                     send macro output to text file
: %R=1!                               display and plot residuals in macros
$OUt 6 80!                                       send output to text file
$Use SYCT V81 V85!                                   fit symmetry models
$PRint / TITL :!
$Use MHCT V81 V85!                          fit marginal homogeneity model
$FINish
```

6.3 Diagonal Models

```
! BRITISH ELECTIONS 1974 (FINGLETON, 1984, P.131)
!  C  LIB    L          PARTIES: C   - CONSERVATIVE
 170   20    3 ! C               LIB - LIBERAL
  22   70   28 ! LIB             L   - LABOUR
   6   12  227 ! L

$UNits 9!                                                      read data
$DAta FREQ!
$PRint 'Load data'!
```

```
$DINput 1!
$Yvariate FREQ!                                              define model
$ERror P!
$FActor OCT 3 FEB 3!
$PRint 'Load SYCT (CONTTAB.glim)'!
$INput 15 SYCT!
$Macro TITL !                                                define title
   BRITISH ELECTION VOTE 1974 (FINGLETON, 1984, P.131)!
$Endmac!
$CAlculate OCT=%GL(3,1)!                         create required variables
: FEB=%GL(3,3)!
$OUt 6 80!                                        send output to text file
$Use SYCT OCT FEB!                                    fit symmetry models
$FINish
```

6.4 Distance and Loyalty Models

```
! BRITISH ELECTIONS 1966 AND 1970 (UPTON, 1978, P.119)
! C   LIB    L    A  1970          PARTIES: C   - CONSERVATIVE
  68    1    1    7 ! C                     LIB - LIBERAL
  12   60    5   10 ! LIB  1966             L   - LABOUR
  12    3   13    2 ! L                     A   - ABSTENTION
   8    2    3    6 ! A

$UNits 16!                                                      read data
$DAta FREQ!
$PRint 'Load data'!
$DINput 1!
$Yvariate FREQ!                                              define model
$ERror P!
$FActor V70 4 V66 4!
$PRint 'Load SYCT (CONTTAB.glim)'!
$INput 15 SYCT!
$Macro TITL !                                                define title
   BRITISH ELECTION VOTE 1966 AND 1970 (UPTON, 1978, P.119)!
$Endmac!
$CAlculate V70=%GL(4,1)!                         create required variables
: V66=%GL(4,4)!
: %R=1!                          display and plot residuals in macros
$OUt 6 80!                                        send output to text file
$Use SYCT V70 V66!                                    fit symmetry models
$FINish
```

APPENDIX III

GLIM MACROS

This macro library is grouped into files by the function of the macros:

test statistics:	CHIT - Chi square probability
variable transformations:	TRAN - conventional constraints
	ORTH - orthogonal polynomials
	IN44 - interactions
contingency tables (general):	MPCT - Markov chain stationarity
	DFCT - correct d.f.
	SMCT - social mobility tables
	BTCT - Bradley-Terry model
	SYCT - symmetry models
	MHCT - marginal homogeneity model
ordinal variables:	L1OV - log multiplicative model
	L2OV - log multiplicative model
	POOV - proportional odds model
	CROV - continuation ratio model
plotting:	RESP - residual plots

The vectors YY1_, YY2_, YY3_, YY4_, YY5_, ZZ1_, ZZ2_, ZZ3_, ZZ4_, ZZ5_, and ZZ9_, as well as any extra ones noted below, are used in these macros and should be avoided when using them. All such vectors end with an underscore as has become the convention for GLIM macros.

The filenames and input channels given below refer to my GLIM macro library on the Amiga and may, of course, be changed to suit the local site.

GLIM is an interpreter, not a compiler, and rereads all characters of every macro every time it executes it. In the interest of efficiency all lower case letters and all unnecessary blank spaces should be eliminated from the following macros in the running version. An example of one such modified macro is given at the end of the appendix.

Macro loaded by $INput 12 from file TESTSTAT.glim.

```
$SUBfile CHIT!
$ECho!
!
!  The macro CHIT calculates the probability level for a given Chi-
!square test.
!  Type $Use CHIT followed by the scalars containing the Chi-square
!value and the d.f.
!  The values of chi-square and d.f. are contained in %DV and %DF
!which may be used with $CAlculate to obtain differences of Chi-
!square.
!  The probability value is returned in %P .
!  Macro used: CHIT
$ECho!
!
$Macro CHIT !
$CAlculate %P=(%2==1)*(2-2*%NP(%SQR(%1)))!calculate 2 special cases
   +(%2==2)*(%EXP(-%1/2))!
   +(%2>2)*(1-%NP(((%1/%2)**(1/3)-1!         calculate general case
   +2/(9*%2))/%SQR(2/(9*%2))))!
$PRint 'Chi2 probability ='%P' for Chi2 ='%1' with'*-4 %2' d.f.' :!
$$Endmac!
$RETurn!
```

Macros loaded by $INput 13 from file DESIGN.glim.

```
$SUBfile TRAN!
$ECho!
!
!  The macro TRAN constructs K-1 dummy contrast variables to
!replace a K level (max 9) factor variable, giving deviations from
!the mean instead of comparisons with the first category of the
!variable.
!  Type $Use TRAN with the factor variable and the list of K-1 new
!dummy variable names.
!  Macros used: TRAN, TRAD
$ECho!
!
$Macro TRAN !
$Argument TRAD %1 %2 %3 %4 %5 %6 %7 %8 %9!
$CAlculate %Z1=%1(1)!                          find number of factor levels
: %Z1=%IF(%1>%Z1,%1,%Z1)!
: %Z2=2!                            set first argument vector to be created
: %Z3=%Z1-1!                                number of vectors to be created
$WHile %Z3 TRAD!                               call macro to create vectors
$$Endmac!
!
```

```
$Macro TRAD !
$CAlculate %%Z2=(%1==%Z2-1)-(%1==%Z1)!       fill vector with values
: %Z2=%Z2+1!                              augment counter to next vector
: %Z3=%Z3-1!                                number of vectors left to do
$$Endmac!
$RETurn!


$SUBfile ORTH!
$ECho!
!
!  The macro ORTH generates linear, quadratic, and cubic orthogonal
!polynomials for any variable, which need not have equally spaced
!factor levels.
!  Type $Use ORTH with the variable name and 3 new variable names
!for the orthogonal polynomials (only 1 or 2 are required if the
!variable has only 2 or 3 levels).
!  See Robson, D.S. (1959) "A simple method for constructing
!orthogonal polynomials when the independent variable is unequally
!spaced." Biometrics 15: 187-191.
$ECho!
!
$Macro ORTH !
$DElete ZZ9_!
$CAlculate %Z3=%CU(1+0*%1)!                  calculate length of vector
: %Z2=%CU(%1)!                              calculate linear polynomial
: %Z2=%Z2/%Z3!
: %2=%1-%Z2!
: %Z4=%CU(%2*%2)!
: %Z4=1/%SQR(%Z4)!
: %2=%2*%Z4!
: %Z4=%CU(%1*%1)!                         calculate quadratic polynomial
: %Z5=%CU(%1*%1*%2)!
: ZZ9_=%1*%1-%Z4/%Z3-%2*%Z5!
: %Z4=%CU(ZZ9_*ZZ9_)!
: %Z1=%Z4<.0001!
$EXit %Z1!                                        stop if only two levels
$CAlculate %Z4=1/%SQR(%Z4)!
: %3=ZZ9_*%Z4!
: %Z4=%CU(%1*%1*%1)!                          calculate cubic polynomial
: %Z5=%CU(%1*%1*%1*%2)!
: %Z6=%CU(%1*%1*%1*%3)!
: ZZ9_=%1*%1*%1-%Z4/%Z3-%2*%Z5-%3*%Z6!
: %Z4=%CU(ZZ9_*ZZ9_)!
: %Z1=%Z4<.0001!
$EXit %Z1!                                      stop if only three levels
$CAlculate %Z4=1/%SQR(%Z4)!
: %4=ZZ9_*%Z4!
$DElete ZZ9_!
$$Endmac!
$RETurn!


$SUBfile IN44!
$ECho!
!
!  The macros IN44 generates all first order interactions between
!two sets of four vectors.
!  Type $Use IN44 with 4+4 variable names, then put #I44 as a term
```

```
!in $Fit. #I44 can then be reapplied in $Fits without retyping
!$Use.
!  When finished, delete macros and interaction variables by typing
!$Use D44, then $DElete D44.
!  Macros used: IN44, I44, D44
!  Extra variables used: RR11 RR12 RR13 RR14 RR21 RR22 RR23 RR24
!RR31 RR32 RR33 RR34 RR41 RR42 RR43 RR34
$ECho!
!
$Macro IN44 !
$CAlculate RR11=%1*%5!          calculate all possible interactions
: RR12=%1*%6!
: RR13=%1*%7!
: RR14=%1*%8!
: RR21=%2*%5!
: RR22=%2*%6!
: RR23=%2*%7!
: RR24=%2*%8!
: RR31=%3*%5!
: RR32=%3*%6!
: RR33=%3*%7!
: RR34=%3*%8!
: RR41=%4*%5!
: RR42=%4*%6!
: RR43=%4*%7!
: RR44=%4*%8!
$$Endmac!
!
$Macro I44 !
   (RR11+RR12+RR13+RR14+RR21+RR22+RR23+RR24+RR31+RR32+RR33+RR34!
   +RR41+RR42+RR43+RR44)!
$Endmac!
!
$Macro D44 !
$DElete IN44 I44 RR11 RR12 RR13 RR14 RR21 RR22 RR23 RR24 RR31 RR32!
   RR33 RR34 RR41 RR42 RR43 RR34!
$$Endmac!
$RETurn!
```

Macros loaded by $INput 15 from file CONTTAB.glim.

```
$SUBfile MPCT!
$ECho!
!
!  The macro MPCT calculates the first order transition
!probabilities for a Markov chain and tests for stationarity.
!  Set up a log linear model as usual with $UNits, $Yvariate,
!$ERror P, $FActor. Define three factor variables: (1) the states
!at the start of a transition, (2) the states at the end of a
!transition, and (3) the time of transition. This third factor
!variable must vary most slowly.
!  Type $Use MPCT with the above three factor variables in order.
!  Macros used: MPCT
$ECho!
!
```

```
$Macro MPCT !
$DElete YY1_ YY2_ ZZ1_ ZZ9_!
$CAlculate %Z2=%1(%NU)!                        calculate size of matrix
: %Z3=%Z2*%Z2!               length of vector containing probabilities
$Variate %Z2 ZZ1_!
: %Z3 YY1_ YY2_!
$CAlculate ZZ1_=0!                                           initialize
: YY1_=0!
: ZZ9_=%GL(%Z3,1)!
: YY1_(ZZ9_)=YY1_(ZZ9_)+%YV!                    calculate marginal sums
: ZZ1_(%1)=ZZ1_(%1)+%YV!
: %Z1=%IF(%1(1)==%1(2),%Z2,1)!         check order of table in vector
: YY2_=%GL(%Z2,%Z1)!
: YY1_=YY1_/ZZ1_(YY2_)!           calculate and print probabilities
$PRint 'First Order Markov Chain' :!
: 'Estimated Stationary Transition Probabilities' :!
$TPrint (S=-1) YY1_ %Z2;%Z2!
$PRint : 'Test for Stationarity'!               fit stationarity model
: $F %1*%2+%1*%3!
$$Endmac!
$RETurn!


$SUBfile DFCT!
$ECho!
!
!  The macro DFCT refits a log linear model to correct the d.f.
!when there are zero cell estimates.
!  Type $CAlculate PW=prior weights (=1 if weights not used).
!  Fit the model, then type $Use DFCT
!  Note that the prior weights (but not the vector PW) have been
!changed after using the macro DFCT and must be reset with $Weight.
!  Macro used: DFCT
$ECho!
!
$Macro DFCT !
$DElete ZZ9_!                 calculate number of valid observations
$CAlculate ZZ9_=((%FV>.001)*((%YV-%FV)**2*PW/%FV/%SC<16)+!
   (%YV/=0)>=1)!
: %T=%CU(ZZ9_)!
: ZZ9_=ZZ9_*PW!
: %T=(%T==%NU)!                                    if all valid, exit
$EXit %T!
$PRint 'Model with corrected df' :!
$Weight ZZ9_!                  refit model eliminating observations
$RECycle 10!
$Fit .!
$CYcle 10!
$$Endmac!
$RETurn!


$SUBfile SMCT!
$ECho!
!
!  The macro SMCT fits Duncan's 5 mobility models to any square 2-
!way mobility table.
!  Set up the table as a log linear model with $UNits, $Yvariate,
!$ERror P, $FActor. Then type $Use SMCT with the father's
```

```
!occupation variable and the son's occupation variable.
!  To write a title at the top of each page, create a macro called
!TITL : $Macro TITL text of title $Endmac.
!  To obtain plots of residuals, type $CAlculate %R=1.
!  See Duncan, O.D. (1979) "How destination depends on origin in
!the occupational mobility table." American Journal of Sociology
!84: 793-803.
!  Macros used: SMCT, TITL, PRC3, CHIT, RESI, RESP, POIS, NORM
!  Extra variable used: PW_, ARE_, SRE_
$ECho!
!
$Macro SMCT !
$DElete ZZ1_ YY1_!
$Use PRC3!
: '1. Independence Model' :!           fit and print out all models
$Fit %1+%2!
$Use CHIT %DV %DF!
$Display E!
$SWitch %R RESI!
$CAlculate %Z2=%1(%NU)!      check if page full; if so, print title
: %Z1=((%Z2>7)+(%R==1)>=1)!
$SWitch %Z1 PRC3!
$PRint '2. Row Effects Model' :!
$CAlculate ZZ1_=%2-1!
$Fit +ZZ1_.%1!
$Use CHIT!
$Display E!
$SWitch %R RESI!
$CAlculate PW_=(%1/=%2)!
$Weight PW_!
$Use PRC3!
$PRint '3. Quasi-independence (Mover-Stayer) Model' :!
$Fit %1+%2!
$Use CHIT!
$Display E!
$SWitch %R RESI!
$CAlculate %Z1=((%K>6)+(%R==1)>=1)!
$SWitch %Z1 PRC3!
$PRint '4. Uniform Association without Diagonal' :!
$CAlculate YY1_=(%1-1)*(%2-1)!
$Fit %1+%2+YY1_!
$Use CHIT!
$Display E!
$SWitch %R RESI!
$Use PRC3!
$PRint '5. Row Effects Model without Diagonal' :!
$Fit %1+%2+ZZ1_.%1!
$Use CHIT!
$Display E!
$SWitch %R RESI!
$$Endmac!
!
$Macro RESI !
$CAlculate %Z1=(%PL+%NU>40)! check if page full; if so, print title
$SWitch %Z1 PRC3!
$Display R!                              display residuals and plot
$Use PRC3!
$Use RESP!
$$Endmac!
```

```
!
$Macro TITL !              blank title to be filled by user if desired
 !
$Endmac!
!
$Macro PRC3 !                          print title at top of new page
$PRint / TITL :!
$$Endmac!
!
$PRint 'Load RESP (GLIMPLOT.glim) and CHIT (TESTSTAT.glim)'$!
$OUt!                                  stop output while loading macros
$TRanscript!
$INput 23 RESP!                                  load required macros
$DElete BINO GAMM OWN!
$INput 12 CHIT!
$OUt 9!                                                restart output
$TRanscript F H I O W!
$RETurn!


$SUBfile BTCT!
$ECho!
!
!  The macro BTCT fits the Bradley-Terry model to any square (KxK)
!table where preferences have been expressed for K items.
!  Set up the log linear model as usual with $UNits, $Yvariate,
!$ERror P, $FActor.
!  Type $Use BTCT with the 2 factor variables, where the first
!varies most quickly.
!  The ranking is given in ascending order as the parameter
!estimates for the second factor variable.
!  See Fienberg, S.E. (1977) The Analysis of Cross-Classified
!Categorical Data. Cambridge: MIT Press, pp.118-121.
!  Macros used: BTCT, CHIT
$ECho!
!
$Macro BTCT !
$DElete YY1_ YY2_ ZZ1_ ZZ9_!
$CAlculate %Z1=%1(%NU)!                           number of items (K)
: ZZ9_=%Z1*(%1-1)+%2!                               calculate indices
: ZZ1_=(%1>%2)!                                            first term
: ZZ1_=ZZ1_*%CU(ZZ1_)!
: YY1_=(%1<%2)!
: YY2_(ZZ9_)=%CU(YY1_(ZZ9_))*YY1_(ZZ9_)!                  second term
: ZZ1_=ZZ1_+YY2_+(%1==%2)!                     factor variable to fit
: ZZ9_=(%1/=%2)!                                     calculate weight
: %Z2=%Z1*(%Z1-1)/2!
$DElete YY1_ YY2_!
$FActor ZZ1_ %Z2!
$Weight ZZ9_!
$Fit %2+ZZ1_!                                               fit model
$Use CHIT %DV %DF!
$Display E!
$$Endmac!
!
$PRint 'Load CHIT (TESTSTAT.glim)'$!
$OUt!                                  stop output while loading macros
$TRanscript!
$INput 12 CHIT!                                   load required macro
```

```
$OUt 9!                                              restart output
$TRanscript F H I O W!
$RETurn!


$SUBfile SYCT!
$ECho!
!
!  The macro SYCT fits symmetry models to square 2-way tables (max
!10x10).
!  Set up the log linear model as usual with $UNits, $Yvariate,
!$ERror P, $FActor.
!  Type $Use SYCT with the 2 factor variables, where the first
!varies most quickly.
!  To write a title at the top of each page, create a macro called
!TITL: $Macro TITL text of title $Endmac.
!  To obtain plots of residuals, type $CAlculate %R=1.
!  See Fingleton, B. (1984) Models of Category Counts. Cambridge:
!Cambridge University Press, pp.130-147.
!  Macros used: SYCT, DIST, PRC3, TITL, CHIT, RESI, RESP, POIS,
!NORM
!  Extra variables used: C1_, C2_, C3, C4_, C5_, C6_, C7_, C8_,
!C9_, PW_, ARE_, SRE_
$ECho!
!
$Macro SYCT !
$DElete ZZ1_ ZZ2_ ZZ3_ ZZ4_ YY1_ YY2_ ZZ9_!
$CAlculate %Z5=%1(%NU)!
: ZZ9_=%Z5*(%1-1)+%2!                          create required vectors
: ZZ1_=(%1>%2)!                                               symmetry
: ZZ1_=ZZ1_*%CU(ZZ1_)!
: ZZ2_=(%1<%2)!                                                loyalty
: YY2_(ZZ9_)=%CU(ZZ2_(ZZ9_))*ZZ2_(ZZ9_)!
: ZZ1_=ZZ1_+YY2_+(%1==%2)!
: %Z4=%Z5*(%Z5-1)/2!
: ZZ2_=(%1==%2)+1!
: ZZ3_=%SQR((%1-%2)**2)+1!                    symmetric minor diagonal
: ZZ4_=%1-%2+1!                              asymmetric minor diagonal
: ZZ4_=%IF(ZZ4_<=0,-ZZ4_+%Z5+1,ZZ4_)!
: %Z6=%Z5*2-1!
: YY1_=(ZZ4_-1)*(%1-%2>0)+1!               minor diagonals (symmetry)
: YY2_=%1!
: ZZ9_=%2!
: %Z2=%Z5!
: C1_=C2_=C3_=C4_=C5_=C6_=C7_=C8_=C9_=0!                 pure distance
$Argument DIST C1_ C2_ C3_ C4_ C5_ C6_ C7_ C8_ C9_!
$WHile %Z2 DIST!
$FActor ZZ1_ %Z4 ZZ2_ 2 ZZ3_ %Z5 ZZ4_ %Z6 YY1_ %Z5!
$Use PRC3!
$PRint '1. Independence Model' :!       fit and print out all models
$Fit %1+%2!
$Use CHIT %DV %DF!
$Display E!
$SWitch %R RESI!
$CAlculate PW_=(%1/=%2)!
$Weight PW_!
$CAlculate %Z3=((%Z5>6)+(%R==1)>=1)!                check if page full;
$SWitch %Z3 PRC3!                                     if so, print title
$PRint '2. Symmetry Model' :!
```

```
$Fit ZZ1_!
$Use CHIT!
$Display E!
$SWitch %R RESI!
$Use PRC3!
$PRint '3. Quasi-symmetry Model' :!
$Fit +%1+%2!
$Use CHIT!
$Display E!
$SWitch %R RESI!
$CAlculate %Z3=((%Z5>4)+(%R==1)>=1)!
$SWitch %Z3 PRC3!
$PRint '4. Minor Diagonals-Symmetry Model' :!
$Fit YY1_+ZZ1_!
$Use CHIT!
$Display E!
$SWitch %R RESI!
$CAlculate PW_=1!
$Use PRC3!
$PRint '5. Main Diagonal (Loyalty) Model' :!
$Fit %1+%2+ZZ2_!
$Use CHIT!
$Display E!
$SWitch %R RESI!
$CAlculate %Z3=((%Z5>6)+(%R==1)>=1)!
$SWitch %Z3 PRC3!
$PRint '6. Symmetric Minor Diagonal Model' :!
$Fit +ZZ3_-ZZ2_!
$Use CHIT!
$Display E!
$SWitch %R RESI!
$Use PRC3!
$PRint '7. Asymmetric Minor Diagonal Model' :!
$Fit +ZZ4_-ZZ3_!
$Use CHIT!
$Display E!
$SWitch %R RESI!
$CAlculate %Z3=((%Z5>5)+(%R==1)>=1)!
$SWitch %Z3 PRC3!
$PRint '8. Pure Distance Model' :!
$Fit %1+%2+C1_+C2_+C3_+C4_+C5_+C6_+C7_+C8_+C9_!
$Use CHIT!
$Display E!
$SWitch %R RESI!
$Use PRC3!
$PRint '9. Loyalty-Distance Model' :!
$Fit +ZZ2_!
$Use CHIT!
$Display E!
$SWitch %R RESI!
$CAlculate PW_=(%1/=%2)!
$SWitch %Z3 PRC3!
$PRint '10. Distance without Main Diagonal' :!
$Fit -ZZ2_!
$Use CHIT!
$Display E!
$SWitch %R RESI!
$ACcuracy 2!                  print out values of all vectors created
$Use PRC3!
```

```
$Look %1 %2 ZZ1_ ZZ2_ ZZ3_ ZZ4_ YY1_!
$CAlculate %Z1=(%Z5>5)!
$PRint!
$SWitch %Z1 PRC3!
$Look %1 %2 C1_ C2_ C3_ C4_ C5_!
$CAlculate %Z1=(%Z5<=6)!
$ACcuracy 4!
$EXit %Z1!
$ACcuracy 2!
$Use PRC3!
$Look %1 %2 C6_ C7_ C8_ C9_!
$ACcuracy 4!
$$Endmac!
!
$Macro DIST !   iterative calculation of vectors for pure distance
!                                                             model
$CAlculate %%Z2=((ZZ9_<=%Z2)*(YY2_>%Z2)+(ZZ9_>%Z2)*(YY2_<=%Z2))!
   *2-1!
: %Z2=%Z2-1!
$$Endmac !
!
$Macro RESI !
$CAlculate %Z1=(%PL+%NU>40)! check if page full; if so, print title
$SWitch %Z1 PRC3!
$Display R!                               display residuals and plot
$Use PRC3!
$Use RESP!
$$Endmac!
!
$Macro TITL !          blank title to be filled by user if desired
 !
$Endmac!
!
$Macro PRC3 !                           print title at top of new page
$PRint / TITL :!
$$Endmac!
!
$PRint 'Load RESP (GLIMPLOT.glim) and CHIT (TESTSTAT.glim)'$!
$OUt!                                stop output while loading macros
$TRanscript!
$INput 23 RESP!                                  load required macros
$DElete BINO GAMM OWN!
$INput 12 CHIT!
$OUt 9!                                                restart output
$TRanscript F H I O W!
$RETurn!


$SUBfile MHCT!
$ECho!
!
!  The macro MHCT fits a marginal homogeneity model to a square 2-
!way table (max. 10x10).
!  Set up the log linear model as usual with $UNits, $Yvariate,
!$ERror P, $FActor. Type $Use MHCT with the 2 factor variables.
!  To $OUtput results to a file, type $CAlculate %O=6.
!  Macros used: MHCT, MARG, ITMH, PRCY, CHIT
!  Extra variables used: C1_, C2_, C3_, C4_, C5_, C6_, C7_, C8_,
!C9_
```

```
$ECho!
!
$Macro MHCT !
$OUt!                          stop output during iterative calculations
$TRanscript!
$DElete YY1_ ZZ1_ ZZ2_ ZZ3_ ZZ4_ ZZ5_!
$CAlculate %Z6=%1(%NU)!                                    size of table
: ZZ1_=%1!                                            initialize vectors
: ZZ2_=%2!
: ZZ5_=ZZ3_=%YV!
: C1_=C2_=C3_=C4_=C5_=C6_=C7_=C8_=C9_=0!
: %Z1=10!                                           number of iterations
: %Z5=0!
: %O=%IF(%O==6,6,9)!                          check where to send output
$Argument ITMH C1_ C2_ C3_ C4_ C5_ C6_ C7_ C8_ C9_!
: MARG %1 %2 %3 %4 %5 %6 %7 %8 %9!
$ERror N!                                                   set up model
$Weight ZZ4_!
$FActor ZZ1_ %Z6 ZZ2_ %Z6!
$WHile %Z1 ITMH!                                                 iterate
$CAlculate YY1_=%GL(%NU,1)!                       calculate unit numbers
: %DV=2*%CU(ZZ3_*%LOG(ZZ3_/ZZ5_))!                   calculate deviance
: %DF=%Z6-1!                                            calculate d.f.
: ZZ3_=(%YV-ZZ5_)/%SQR(ZZ5_)!                       calculate residuals
$OUt %O!                                                  restart output
$TRanscript F H I O W!
$PRint 'Marginal Homogeneity Model' :!                       print model
: 'scaled deviance ='%DV' at cycle '*-2 %Z5!
: '         d.f. = '*-2 %DF :!
$SWitch %Z4 PRCY!
$Use CHIT %DV %DF!
$Display E!
$PRint '   unit observed  fitted  residual'!
$Look (S=-1) YY1_ %YV ZZ5_ ZZ3_!
$$Endmac!
!
$Macro ITMH !                                          iterative fitting
$CAlculate ZZ4_=1/ZZ5_!
: %Z5=%Z5+1!
: %Z2=%Z6-1!
$WHile %Z2 MARG!
$Fit C1_+C2_+C3_+C4_+C5_+C6_+C7_+C8_+C9_-%GM!
$CAlculate ZZ5_=ZZ3_-%FV!
: %Z3=%DV-%Z2!
: %Z2=%DV!
: %Z1=%IF(%Z4=(%Z3*%Z3>.0001),%Z1-1,0)!
$$Endmac!
!
$Macro MARG !                                          calculate vectors
$CAlculate %%Z2=((ZZ1_==%Z2)-(ZZ2_==%Z2))*ZZ5_!
: %Z2=%Z2-1!
$$Endmac!
!
$Macro PRCY !                   message to print if no convergence
$PRint '    (no convergence yet)' :!
$$Endmac!
!
$PRint 'Load CHIT (TESTSTAT.glim)'$!
$OUt!                                stop output while loading macros
```

```
$TRanscript!
$INput 12 CHIT!                                   load required macro
$OUt 9!                                                restart output
$TRanscript F H I O W!
$RETurn!
```

Macros loaded by $INput 16 from file ORDVAR.glim.

```
$SUBfile L1OV!
$ECho!
!
!  The macro L1OV calculates the appropriate scale for the
!relationship  between a (combination of) nominal and an ordinal
!variable in a log-linear model.
!  Set up the model as usual with $UNits, $Yvariate, $ERror P,
!$FActor. The two variables concerned must both be declared in
!$FActor. Type $Use L1OV with the two variable names (second one
!ordinal).
!  The scale is returned in a new quantitative variable, ZZ1_,
!which may be used in subsequent $Fits, for example, if a combined
!nominal variable is fitted as separate variables.
!  To obtain plots of residuals, type $CAlculate %R=1.
!  To $OUtput results to a file, type $CAlculate %O=6.
!  To write a title at the top of each page, create a macro called
!TITL: $Macro TITL text of title $Endmac.
!  See Anderson, J.A. (1984) "Regression and ordered categorical
!variables." Journal of the Royal Statistical Society (B) 46: 1-30.
!  Macros used: L1OV, PRC1, PRC3, ITER, CHIT, RESI, RESP, POIS,
!NORM, TITL
!  Extra variables used: PW_, ARE_, SRE_
$ECho!
!
$Macro L1OV !
$DElete YY1_ YY2_ YY3_ YY4_ ZZ1_ ZZ2_ ZZ3_ ZZ4_!
$CAlculate %Z8=%1(1)!                    calculate dimensions of table
: %Z8=%IF(%1>%Z8,%1,%Z)!
: %Z9=%2(1)!
: %Z9=%IF(%2>%Z9,%2,%Z9)!
: %Z4=%Z6=0!                                       initialize counters
: %Z2=10!
: %O=%IF(%O==6,6,9)!                          check where to send output
: YY4_=2*%1-%Z8-1!                          calculate 2 linear variables
: ZZ4_=2*%2-%Z9-1!
$OUt %O!
$PRint / TITL :!
: 'Independence Model' :!
$Fit %1+%2!
$Use CHIT %DV %DF!
$Display E!
$SWitch %R RESI!
$CAlculate %Z1=((%PL>16)+(%R==1)>=1)!              check if page full;
$SWitch %Z1 PRC3!                                    if so, print title
$PRint 'Linear Effects Model' :!
$Fit %1+%2+%1.ZZ4_!
$Use CHIT!
$Display E!
```

```
$SWitch %R RESI!
$OUt!                          stop output during iterative calculations
$TRanscript!
$Variate %Z8 YY2_ YY3_!
: %Z9 ZZ2_ ZZ3_!
$CAlculate YY1_=%1!
: YY3_=%GL(%Z8,1)+%Z8!
: ZZ3_=%GL(%Z9,1)+%Z9!
$Argument ITER %1 %2!
$WHile %Z2 ITER!
$Fit %1+%2+%1.ZZ1_!                                     final fit for scale
$CAlculate %Z4=%DF-%Z9+2!                                              d.f.
: %Z3=ZZ2_(1)!                  standardize scale to lie between 0 and 1
: %Z7=ZZ2_(%Z9)-%Z3!
: ZZ2_=(ZZ2_-%Z3)/%Z7!
$OUt %O!                                                     restart output
$TRanscript F H I O W!
$PRint / TITL :!
: 'Scale for ordinal variable'!
: ZZ2_ :!
: 'Log Multiplicative Model' :!                            print out model
: 'scaled deviance = ' *4 %DV ' at cycle' *-2 %Z6!
: '             d.f. = ' *-2 %Z4 :!
$SWitch %Z5 PRC1!
$Use CHIT!
$Display E!
$SWitch %R RESI!
$DElete YY1_ YY2_ YY3_ YY4_ ZZ2_ ZZ3_ ZZ4_!
$$Endmac!
!
$Macro ITER !                                        iterative fitting macro
$CAlculate %Z6=%Z6+1!
: %Z5=%DV!
$Fit YY1_+%2+YY1_.%2!                                  estimate first scale
$EXTract %PE!
$CAlculate %Z3=%PE(2)!
: ZZ2_=(ZZ3_/=%Z9+1)*%PE(ZZ3_)+%Z3!
: ZZ1_=(%2/=1)*%PE(%2+%Z9)+%Z3!
: %Z3=ZZ2_(1)!
: %Z7=ZZ2_(%Z9)-%Z3!
: ZZ1_=(ZZ1_-%Z3)/%Z7!
$Fit ZZ1_+%1.ZZ1_+%1!                                 estimate second scale
$EXTract %PE!
$CAlculate %Z3=%PE(2)!
: YY2_=(YY3_/=%Z8+1)*%PE(YY3_)+%Z3!
: YY1_=(%1/=1)*%PE(%1+%Z8)+%Z3!
: %Z3=YY2_(1)!
: %Z7=YY2_(%Z8)-%Z3!
: YY1_=(YY1_-%Z3)/%Z7!
: %Z5=(%Z5-%DV)/%DV!                                   test for convergence
: %Z2=%IF(%Z5=(%Z5*%Z5>.00001),%Z2-1,0)!
$$Endmac!
!
$Macro RESI !
$CAlculate %Z1=(%PL+%NU>40)! check if page full; if so, print title
$SWitch %Z1 PRC3!
$Display R!                                    display residuals and plot
$Use PRC3!
$Use RESP!
```

```
$$Endmac!
!
$Macro TITL !             blank title to be filled by user if desired
 !
$Endmac!
!
$Macro PRC1 !                        message to print if no convergence
$PRint '    (no convergence yet)' :!
$$Endmac!
!
$Macro PRC3 !                           print title at top of new page
$PRint / TITL :!
$$Endmac!
!
$PRint 'Load RESP (GLIMPLOT.glim) and CHIT (TESTSTAT.glim)'$!
$OUt!                                stop output while loading macros
$TRanscript!
$INput 23 RESP!                                  load required macros
$DElete BINO GAMM OWN!
$INput 12 CHIT!
$OUt 9!                                                restart output
$TRanscript F H I O W!
$RETurn!


$SUBfile L2OV!
$ECho!
!
!  The macro L2OV calculates the appropriate scores for the
!relationship between two ordinal variables in a log-linear model.
!  Set up the model as usual with $UNits, $Yvariate, $ERror P,
!$FActor. The two variables concerned must both be declared in
!$FActor. Type $Use L2OV with the two variable names.
!  The scales are returned in two new quantitative variables, YY1_
!and ZZ1_ .
!  To obtain plots of residuals, type $CAlculate %R=1.
!  To $OUtput results to a file, type $CAlculate %O=6.
!  To write a title at the top of each page, create a macro called
!TITL: $Macro TITL text of title $Endmac
!  See Goodman, L.A. (1979) "Simple models for the analysis of
!association in cross-classifications having ordered categories."
!Journal of the American Statistical Association 74: 537-552 and
!Goodman, L.A. (1981) "Association models and canonical correlation
!in the analysis of cross-classifications having ordered
!categories." Journal of the American Statistical Association 76:
!320-334.
!  MACROS USED: L2OV, PRC1_, PRC3, ITER, RESI, RESP, POIS, NORM,
!CHIT, TITL
!  Extra variables used: PW_, ARE_, SRE_
!  Extra scalars used: %M, %N, %P, %Q, %S, %T, %U, %V, %X, %Y, %Z
$ECho!
!
$Macro L2OV !
$DElete YY1_ YY2_ YY3_ YY4_ ZZ1_ ZZ2_ ZZ3_ ZZ4_ ZZ9_!
$CAlculate %Z8=%1(1)!                   calculate dimensions of table
: %Z8=%IF(%1>%Z8,%1,%Z8)!
: %Z9=%2(1)!
: %Z9=%IF(%2>%Z9,%2,%Z9)!
: %Z4=%Z6=0!                                       initialize counters
```

```
: %Z2=10!
: %O=%IF(%O==6,6,9)!                          check where to send output
: YY4_=2*%1-%Z8-1!                            calculate 2 linear variables
: ZZ4_=2*%2-%Z9-1!
$OUt %O!
$PRint / TITL :!                                        print out all models
: 'Independence Model' :!
$Fit %1+%2!
$Display E!
$SWitch %R RESI!
$CAlculate %Y=%DV!
: %X=%DF!
: %Z1=((%PL>20)+(%R==1)>=1)!                          check if page full;
$SWitch %Z1 PRC3!                                      if so, print title
$PRint 'Linear Effects Model' :!
$CAlculate ZZ9_=YY4_*ZZ4_!
$Fit %1+%2+ZZ9_!
$Display E!
$SWitch %R RESI!
$OUt!                    stop output during iterative calculations
$TRanscript!
$CAlculate YY1_=%1!
: %M=%DV!
: %N=%DF!
$Variate %Z8 YY2_ YY3_!
: %Z9 ZZ2_ ZZ3_!
$CAlculate YY3_=%GL(%Z8,1)+%Z8!
: ZZ3_=%GL(%Z9,1)+%Z9!
$Argument ITER %1 %2!
$WHile %Z2 ITER!
$OUt %O!
$TRanscript F H I O W!                                    restart output
$PRint / TITL :!
: 'Column (K) Effect Model' :!
$Fit %1+%2+ZZ4_.%1!
$Display E!
$SWitch %R RESI!
$CAlculate %Z1=((2*%PL-%Z8+%Z9>40)+(%R==1)>=1)!
: %Z=%DV!
: %Q=%DF!
$SWitch %Z1 PRC3!
$PRint 'Row (I) Effect Model' :!
$Fit %1+%2+YY4_.%2!
$Display E!
$SWitch %R RESI!
$CAlculate %V=%DV!
: %S=%DF!
$PRint / TITL :!
: 'Row and Column Effect Model (1)' :!
$Fit %1+%2+%1.ZZ4_+YY4_.%2!
$Display E!
$SWitch %R RESI!
$OUt!                                 stop output during during fit
$TRanscript!                                         and Chi-square
$CAlculate ZZ9_=YY1_*ZZ1_!
: %T=%DV!
: %U=%DF!
$Fit %1+%2+ZZ9_!                                   final fit for scale
$CAlculate %Z4=%DF-%Z9-%Z8+4!                                       d.f.
```

```
: %Z3=ZZ2_(1)!              standardize scale to lie between 0 and 1
: %Z7=ZZ2_(%Z9)-%Z3!
: ZZ2_=(ZZ2_-%Z3)/%Z7!
: %Z3=YY2_(1)!
: %Z7=YY2_(%Z8)-%Z3!
: YY2_=(YY2_-%Z3)/%Z7!
: %X=%X-%N!                                      calculate Chi squares
: %Y=%Y-%M!
: %Q=%N-%Q!
: %Z=%M-%Z!
: %S=%N-%S!
: %V=%M-%V!
$Use CHIT %Y %X!
$CAlculate %Z2=%P!
$Use CHIT %Z %Q!
$CAlculate %Z3=%P!
$Use CHIT %V %S!
$CAlculate %M=%P!
$Use CHIT %T %U!
$CAlculate %N=%P!
$Use CHIT %DV %Z4!
$OUt %O!
$TRanscript F H I O W!                                  restart output
$PRint / TITL :!
: 'Row and Column Effect Model (2)' :!                 print out model
: 'Scale for First Ordinal Variable'!
: YY2_ :!
: 'Scale for Second Ordinal Variable' :!
: ZZ2_ :!
: 'scaled deviance = ' *4 %DV ' at cycle' *-2 %Z6!
: '           d.f. = ' *-2 %Z4:!
$SWitch %Z5 PRC1!
$Display E!
$SWitch %R RESI!
$CAlculate %Z1=((%PL>36)+(%R==1)>=1)!
$SWitch %Z1 PRC3!
$PRint 'Analysis of Association Table' :!
: '    Effect            Chi2   df   Prob'!
: 'General Effect     ' *4 %Y *-3 %X *4 %Z2!
: 'Column Effect      ' *4 %Z *-3 %Q *4 %Z3!
: 'Row Effect         ' *4 %V *-3 %S *4 %M!
: 'Other Effects(1) ' *4 %T *-3 %U *4 %N!
: 'Other Effects(2) ' *4 %DV *-3 %Z4 *4 %P!
$DElete YY2_ YY3_ YY4_ ZZ2_ ZZ3_ ZZ4_ ZZ9_!
$$Endmac!
!
$Macro ITER !                               iterative fitting macro
$CAlculate %Z6=%Z6+1!
: %Z5=%DV!
$Fit YY1_+%2+YY1_.%2!                          estimate first scale
$EXTract %PE!
$CAlculate %Z3=%PE(2)!
: ZZ2_=(ZZ3_/=%Z9+1)*%PE(ZZ3_)+%Z3!
: ZZ1_=(%2/=1)*%PE(%2+%Z9)+%Z3!
: %Z3=ZZ2_(1)!
: %Z7=ZZ2_(%Z9)-%Z3!
: ZZ1_=(ZZ1_-%Z3)/%Z7!
$Fit ZZ1_+%1.ZZ1_+%1!                         estimate second scale
$EXTract %PE!
```

```
$CAlculate %Z3=%PE(2)!
: YY2_=(YY3_/=%Z8+1)*%PE(YY3_)+%Z3!
: YY1_=(%1/=1)*%PE(%1+%Z8)+%Z3!
: %Z3=YY2_(1)!
: %Z7=YY2_(%Z8)-%Z3!
: YY1_=(YY1_-%Z3)/%Z7!
: %Z5=(%Z5-%DV)/%DV!                              test for convergence
: %Z2=%IF(%Z5=(%Z5*%Z5>.00001),%Z2-1,0)!
$$Endmac!
!
$Macro RESI !
$CAlculate %Z1=(%PL+%NU>40)! check if page full; if so, print title
$SWitch %Z1 PRC3!
$Display R!                                 display residuals and plot
$Use PRC3!
$Use RESP!
$$Endmac!
!
$Macro TITL !             blank title to be filled by user if desired
 !
$Endmac!
!
$Macro PRC1 !                         message to print if no convergence
$PRint '    (no convergence yet)' :!
$$Endmac!
!
$Macro PRC3 !                             print title at top of new page
$PRint / TITL :!
$$Endmac!
!
$PRint 'Load RESP (GLIMPLOT.glim) and CHIT (TESTSTAT.glim)'$!
$OUt!                                 stop output while loading macros
$TRanscript!
$INput 23 RESP!                                   load required macros
$DElete BINO GAMM OWN!
$INput 12 CHIT!
$OUt 9!                                                 restart output
$TRanscript F H I O W!
$RETurn!


$SUBfile POOV!
$ECho!
!
!  The macro POOV fits a proportional odds model for an ordinal
!dependent variable with grouped frequency data.
!  Type $CAlculate %N=number of independent variables, %K=number
!of categories of the dependent variable (%N+%K <= 10), and
!%L=number of lines in the table (table size = %Lx%K). Type $Use
!POOV with the names of the frequency vector and up to 7
!independent variables
!  The ordinal dependent variable must vary most quickly in the
!frequency vector. All independent variables must be continuous or
!binary ($FActor cannot be used; instead, apply the macro TRAN).
!  The first %K-1 parameter estimates refer to the odds for
!categories of the dependent variable, the last %N to the
!independent variable.
!  To $OUtput results to a file, type $CAlculate %O=6.
!  To write a title at the top of each page, create a macro called
```

```
!TITL: $Macro TITL text of title $Endmac
!  See Hutchison, D. (1985) "Ordinal variable regression using the
!McCullagh (proportional odds) model." GLIM Newsletter 9: 9-17.
!  Macros used: POOV, INDV, IND1, STEP, CM, MEXT, ETML, WMU, PARA,
!FV, DR, VA, DI, PRC3, TITL, CHIT
!  Extra variables used: C1_, C2_, C3_, C4_, C5_, C6_, C7_, C8_,
!C9_
$ECho!
!
$Macro POOV !
$OUt!
$TRanscript!
$DElete YY2_ YY3_ YY5_ ZZ1_ ZZ2_ ZZ5_ ZZ9_ C1_ C2_ C3_ C4_ C5_ C6_
    C7_ C8_ C9_ !
$CAlculate %Z7=9!
: %Z2=%K*%L!
: %O=%IF(%O==6,6,9)!                        check where to send output
$Argument STEP %1!
: INDV %2 %3 %4 %5 %6 %7 %8 %9!
: IND1 C1_ C2_ C3_ C4_ C5_ C6_ C7_ C8_ C9_!
$CAlculate YY5_=%GL(%L,%K)!    create indices to manipulate vectors
: %Z4=(%K-1)*%L!
$Variate %L ZZ5_!
: %NU ZZ9_!
$UNits %Z4!
$CAlculate ZZ5_=0!                                    calculate totals
: ZZ5_(YY5_)=ZZ5_(YY5_)+%1!
: ZZ2_=%GL(%L,1)!
: %Z1=%K-1!                             calculate dependent variable
: YY2_=0!
$WHile %Z1 STEP!
$DElete YY5_ ZZ9_!
$CAlculate %Z4=%NU*%Z7!  initialize to create independent variables
$Variate %Z4 ZZ1_ YY3_ YY5_ ZZ9_!
$CAlculate YY5_=%GL(%Z7,1)!
: YY3_=(YY5_==%GL(%K-1,%Z7*%L))!
: ZZ9_=(%GL(%L,%Z7)-1)*%K+%GL(%K,%L*%Z7)!
: %Z8=%N!
$WHile %Z8 INDV!          set up independent variables in new vectors
$Variate %NU C1_ C2_ C3_ C4_ C5_ C6_ C7_ C8_ C9_!
$CAlculate %Z8=%Z7!
$WHile %Z8 IND1!
$DElete YY3_ YY5_ ZZ1_ ZZ9_!
$CAlculate YY3_=ZZ5_(ZZ2_)!                    transform totals vector
: ZZ1_=%GL(%K-1,%L)!
$ERror B YY3_!                                set up approximate model
$Yvariate YY2_!
$OUt %O!
$TRanscript F H I O W!
$PRint 'Proportional Odds Model' :!           print transformed table
: '       R         N        ZZ1_         ZZ2_'!
$Look (S=-1) YY2_ YY3_ ZZ1_ ZZ2_!
$DElete ZZ1_ ZZ2_!
$PRint : 'Approximate Analysis' :!
$Fit C1_+C2_+C3_+C4_+C5_+C6_+C7_+C8_+C9_-%GM! fit approximate model
$Use CHIT %DV %DF!                            for initial estimates
$Display E!
$OUt!
$TRanscript!
```

```
$EXTract %PE!
$DElete YY2_ YY3_ C1_ C2_ C3_ C4_ C5_ C6_ C7_ C8_ C9_!
$UNits %Z2!
$CAlculate %Z4=%Z2*%Z7!  initialize to create independent variables
$Variate %Z4 YY3_ YY5_ ZZ1_ ZZ9_!
$CAlculate YY5_=%GL(%Z7,1)!
: YY3_=(YY5_==%GL(%K,%Z7))*(YY5_<%K)!
: ZZ9_=%GL(%NU,%Z7)!
: %Z8=%N!
$WHile %Z8 INDV!        set up independent variables in new vectors
$DElete ZZ9_!
$CAlculate ZZ9_=%GL(%L,%K)!
: YY2_=ZZ5_(ZZ9_)!
: ZZ9_=%GL(%NU,1)!
$DElete YY5_ ZZ1_ ZZ5_!
$Yvariate %1!                                     set up exact model
$OWn FV DR VA DI!
$SCale 1!
$Argument FV C1_ C2_ C3_ C4_ C5_ C6_ C7_ C8_ C9_!
: WMU %2 %3 %4 %5 %6 %7 %8 %9 %1!
$CAlculate %LP=C1_=C2_=C3_=C4_=C5_=C6_=C7_=C8_=C9_=0!
: %Z1=(%L*(%K-1)>=8)!
$OUt %O!
$TRanscript F H I O W!
$SWitch %Z1 PRC3!
$PRint 'Exact Analysis' :!                          fit exact model
$Fit C1_+C2_+C3_+C4_+C5_+C6_+C7_+C8_+C9_-%GM!
$Use CHIT!
$Display E!
$OUt 9!
$$Endmac!
!
$Macro STEP !                           fills new vector with values
$CAlculate ZZ9_=(%GL(%K,1)<=%K-%Z1)*(%GL(%L,%K)+(%K-1-%Z1)*%L)!
: YY2_(ZZ9_)=YY2_(ZZ9_)+%1!
: %Z1=%Z1-1!
$$Endmac!
!
$Macro INDV !         first step to create new independent variables
$CAlculate ZZ1_=(YY5_==%K+%Z8-1)*ZZ9_!
: YY3_=YY3_+%%Z8(ZZ1_)!
: %Z8=%Z8-1!
$$Endmac!
!
$Macro IND1 !        second step to create new independent variables
$CAlculate ZZ1_=(YY5_==%Z8)*%GL(%NU,%Z7)!
: %%Z8(ZZ1_)=YY3_!
: %Z8=%Z8-1!
$$Endmac!
!
$Macro FV !                                    fitted values for own
$CAlculate %Z3=(%PL/=0)!
: %Z6=%Z7!
: ZZ2_=40*(%GL(%K,1)==%K)!
$SWitch %Z3 MEXT!       skip extraction of estimates the first time
$WHile %Z6 ETML!
$CAlculate ZZ5_=%EXP(ZZ2_)/(1+%EXP(ZZ2_))!        transform previous
!                                                          estimates
: YY5_=ZZ5_/(1+%EXP(ZZ2_))!
```

```
: %Z5=1!
$Use CM ZZ5_ %FV!                         calculate fitted values vector
$CAlculate %Z5=2!
$Use CM ZZ2_ %LP!                      calculate linear predictor vector
$CAlculate %Z6=%Z7!
$Use WMU!
$$Endmac!
!
$Macro DR !                          calculate deta by dgamma for own
$CAlculate %DR=1!
$Endmac!
!
$Macro VA !                        calculate variance function for own
$CAlculate %VA=%FV!
$Endmac!
!
$Macro DI !                      calculate increase in deviance for own
$CAlculate %DI=2*(%YV*%LOG((%YV+(%YV==0))/%FV)-(%YV-%FV))!
$Endmac!
!
$Macro ETML !                                            calculate eta
$CAlculate ZZ2_=ZZ2_+%PE(%Z6)*YY3_(%Z7*(ZZ9_-1)+%Z6)!
: %Z6=%Z6-1!
$$Endmac!
!
$Macro MEXT !                               obtain parameter estimates
$EXTract %PE!
$$Endmac!
!
$Macro CM !                   set up fitted values and linear predictor
$CAlculate %1=((%Z5==1)+(%Z5==2)*YY5_)*%1!
: %2=(%1-%1(ZZ9_-1)*(%GL(%K,1)/=1))*YY2_!
$$Endmac!
!
$Macro WMU !                       calculate new parameter estimates
$CAlculate ZZ1_=YY3_((ZZ9_-1)*%Z7+(%Z7-%Z6+1))!
$Use CM ZZ1_ %9!
$CAlculate %Z8=((%Z6=%Z6-1)>0)!
$SWitch %Z8 WMU!
$$Endmac!
!
$Macro TITL !             blank title to be filled by user if desired
 !
$Endmac!
!
$Macro PRC3 !                           print title at top of new page
$PRint / TITL :!
$$Endmac!
!
$PRint 'Load CHIT (TESTSTAT.glim)'$!
$OUt!                                 stop output while loading macros
$TRanscript!
$INput 12 CHIT!                                    load required macro
$OUt 9!                                                 restart output
$TRanscript F H I O W!
$RETurn!
```

```
$SUBfile CROV!
$ECho!
!
!  The macro CROV fits a continuation ratio model for an ordinal
!dependent variable with grouped frequency data.
!  Set $UNits and read frequency data as for a log-linear model
!into a vector, where the dependent ordinal variable varies most
!quickly. Use $CAlculate to set %K=number of categories of ordinal
!variable and %L=product of number of categories of all independent
!variables. Then type $Use CROV with the frequency vector.
!  ZZ1_ is the vector defining the series of continuation ratio
!models. ZZ2_ is the vector defining the combination of independent
!variables. ZZ2_ may be replaced by these variables, if they are
!redefined by %GL and $FActor. The model may then be refitted with
!them and ZZ1_.
!  To $OUtput results to a file, type $CAlculate %O=6.
!  To write a title at the top of each page, create a macro called
!TITL: $Macro TITL text of title $Endmac
!  Macros used: CROV, PRC3, STEP, TITL, CHIT
!  Extra variables used: PW_
$ECho!
!
$Macro CROV !
$OUt!
$TRanscript!
$DElete PW_ ZZ1_ ZZ2_ ZZ3_ ZZ4_ ZZ5_ YY1_ YY2_ YY3_!
$CAlculate %O=%IF(%O==6,6,9)!            check where to send output
: YY1_=%GL(%L,%K)!                                     create indices
: ZZ3_=%GL(%K,1)!
: %NU=(%K-1)*%L!
: ZZ4_=0!
: %Z1=%K-1!
$UNits %NU!
$ERror B YY2_!                                         set up model
$Yvariate YY3_!
$FActor ZZ2_ %L ZZ1_ %Z1!
$Weight PW_!
$CAlculate ZZ2_=%GL(%L,1)!                         create new table
: ZZ1_=%GL(%Z1,%L)!
: YY2_=0!
: YY3_=0!
: PW_=1!
$Argument STEP %1!
$WHile %Z1 STEP!
$CAlculate YY3_=YY2_-YY3_!
$DElete YY1_ ZZ3_ ZZ4_!
$OUt %O!
$TRanscript F H I O W!
$PRint 'Continuation Ratio Model' :!                    print table
: '       R          N          ZZ1_        ZZ2_'!
$Look (S=-1) YY3_ YY2_ ZZ1_ ZZ2_!
$CAlculate %Z1=(%NU+%PL>=40)!
$PRint!
$SWitch %Z1 PRC3!
$Fit ZZ1_+ZZ2_!                                           fit model
$Use CHIT %DV %DF!
$Display E!
$$Endmac!
!
```

```
$Macro STEP !                                         creates new vectors
$CAlculate ZZ4_=(ZZ3_<=%K-%Z1+1)*(YY1_+%L*(%K-%Z1-1))!
: YY2_(ZZ4_)=YY2_(ZZ4_)+%1!
: ZZ4_=ZZ4_*(ZZ3_==%K-%Z1+1)!
: YY3_(ZZ4_)=YY3_(ZZ4_)+%1!
: %Z1=%Z1-1!
$$Endmac!
!
$Macro TITL !          blank title to be filled by user if desired
 !
$Endmac!
!
$Macro PRC3 !                          print title at top of new page
$PRint / TITL :!
$$Endmac!
!
$PRint 'Load CHIT (TESTSTAT.glim)'$!
$OUt!                                 stop output while loading macro
$TRanscript!
$INput 12 CHIT!                                  load required macro
$OUt 9!                                               restart output
$TRanscript F H I O W!
$RETurn!
```

Macro loaded by $INput 23 from file GLIMPLOT.glim.

```
$SUBfile RESP!
$ECho!
!
!  The macro RESP plots standardized and adjusted residuals against
!the normal order statistic and the score test coefficient of
!sensitivity for any GLM.
!  Note that prior weights are defined as 1 with $Weight when this
!macro is loaded.
!  After fitting a model, type $Use RESP.
!  To $OUtput results to a file, type $CAlculate %O=6.
!  See Gilchrist, R. (1982) "GLIM syntax for adjusted residuals."
!GLIM Newsletter 6: 64-65.
!  Macros used: RESP, NORM, BINO, POIS, GAMM, OWN
!  Extra variables used: ARE_, SRE_, PW_
$ECho!
!
$Macro RESP !
$EXTract %VL!
$DElete ARE_ SRE_ YY5_ ZZ5_!
$CAlculate %O=%IF(%O==6,6,9)!               check where to send output
$OUt %O!
$TRanscript F H I O W!
$SWitch %ERR NORM POIS BINO GAMM OWN OWN OWN OWN OWN!
$OUt!
$TRanscript!
$CAlculate ZZ5_=%WT*%VL/%SC!
: ARE_=(%YV-%FV)*%SQR(%PW/YY5_/%SC/(1-ZZ5_))!     adjusted residuals
: SRE_=ARE_*ARE_*ZZ5_/(ZZ5_-1)!             coefficient of sensitivity
: YY5_=%GL(%NU,1)!                                  observation number
: %RE=(%PW/=0)!             eliminate observations with zero weight
```

```
$OUt %O 80!
$TRanscript F H I O W!
$PRint 'Score Test Coefficient of Sensitivity' :!
$Plot SRE_ YY5_!
$OUt!
$TRanscript!
$CAlculate SRE_=ARE_*%SQR(1-ZZ5_)!              standardized residuals
: YY5_=%ND((%GL(%NU,1)-.5)/%NU)!                normal order statistic
$Sort %RE %RE ARE_!
: ARE_!
: SRE_!
$OUt %O 80!
$TRanscript F H I O W!
$PRint : 'Residual Plot' :!
$Plot ARE_ SRE_ YY5_ YY5_!
$PRint : 'Points Y represent 45 line'!
$DElete ARE_ SRE_ YY5_ ZZ5_ %RE!
$$Endmac!
!
$Macro NORM !    macros to calculate appropriate variance function
$CAlculate YY5_=1!
$PRint 'Normal Residuals' :!
$$Endmac!
!
$Macro POIS !
$CAlculate YY5_=%FV!
$PRint 'Poisson Residuals' :!
$$Endmac!
!
$Macro BINO !
$CAlculate YY5_=%FV*(%BD-%FV)/%BD!
$PRint 'Binomial Residuals' :!
$$Endmac!
!
$Macro GAMM !
$CAlculate YY5_=%FV*%FV!
$PRint 'Gamma Residuals' :!
$$Endmac!
!
$Macro OWN !
$CAlculate YY5_=%VA!
$PRint 'Own Residuals' :!
$$Endmac!
!
$CAlculate PW_=1!     weight must be defined to calculate residuals
$Weight PW_!           so give default value which may subsequently
$RETurn!                                             be changed by user
```

An example of a macro (L1OV) with all unnecessary characters eliminated in the interest of efficiency:

```
$M L1OV $DE YY1_ YY2_ YY3_ YY4_ ZZ1_ ZZ2_ ZZ3_ ZZ4_$CA %Z8=%1(1)!
:%Z8=%IF(%1>%Z8,%1,%Z):%Z9=%2(1):%Z9=%IF(%2>%Z9,%2,%Z9):%Z4=%Z6=0!
:%Z2=10:%O=%IF(%O==6,6,9):YY4_=2*%1-%Z8-1:ZZ4_=2*%2-%Z9-1$OU %O!
$PR/TITL::'Independence   Model':$F %1+%2$U CHIT %DV %DF$D E!
```

```
$SW %R RESI$CA %Z1=((%PL>16)+(%R==1)>=1)$SW %Z1 PRC3!
$PR'Linear  Effects  Model':$F %1+%2+%1.ZZ4_$U CHIT$D E$SW %R RESI
$OU$TR$V %Z8 YY2_ YY3_:%Z9 ZZ2_ ZZ3_$CA YY1_=%1!
:YY3_=%GL(%Z8,1)+%Z8:ZZ3_=%GL(%Z9,1)+%Z9$A ITER %1 %2!
$WH %Z2 ITER$F %1+%2+%1.ZZ1_$CA %Z4=%DF-%Z9+2:%Z3=ZZ2_(1)!
:%Z7=ZZ2_(%Z9)-%Z3:ZZ2_=(ZZ2_-%Z3)/%Z7$OU %O$TR F H I O W!
$PR/TITL::'Scale for ordinal variable':ZZ2_ :!
:'Log Multiplicative Model':!
:'scaled deviance = ' *4 %DV ' at cycle' *-2 %Z6!
:'         d.f. = ' *-2 %Z4 :$SW %Z5 PRC1$U CHIT$D E$SW %R RESI!
$DE YY1_ YY2_ YY3_ YY4_ ZZ2_ ZZ3_ ZZ4_$$E!
$M ITER $CA %Z6=%Z6+1:%Z5=%DV$F YY1_+%2+YY1_.%2$EXT %PE!
$CA %Z3=%PE(2):ZZ2_=(ZZ3_/=%Z9+1)*%PE(ZZ3_)+%Z3!
:ZZ1_=(%2/=1)*%PE(%2+%Z9)+%Z3:%Z3=ZZ2_(1):%Z7=ZZ2_(%Z9)-%Z3!
:ZZ1_=(ZZ1_-%Z3)/%Z7$F ZZ1_+%1.ZZ1_+%1$EXT %PE$CA %Z3=%PE(2)!
:YY2_=(YY3_/=%Z8+1)*%PE(YY3_)+%Z3:YY1_=(%1/=1)*%PE(%1+%Z8)+%Z3!
:%Z3=YY2_(1):%Z7=YY2_(%Z8)-%Z3:YY1_=(YY1_-%Z3)/%Z7!
:%Z5=(%Z5-%DV)/%DV:%Z2=%IF(%Z5=(%Z5*%Z5>.00001),%Z2-1,0)$$E!
$M RESI $CA %Z1=(%PL+%NU>40)$SW %Z1 PRC3$D R$U PRC3$U RESP$$E!
$M TITL  $E!
$M PRC1 $PR '    (no convergence yet)' :$$E!
$M PRC3 $PR / TITL :$$E!
$PR'Load RESP (GLIMPLOT.glim) and CHIT (TESTSTAT.glim)'$OU$TR!
$IN 23 RESP$DE BINO GAMM OWN$IN 12 CHIT$OU 9$TR F H I O W!
$RET!
```

REFERENCES

Agresti, A. (1984) **Analysis of Ordinal Categorical Data.** New York: Wiley.

Aickin, M. (1983) **Linear Statistical Analysis of Discrete Data.** New York: Wiley.

Aitkin, M., Anderson, D., Francis, B. & Hinde, J. (1989) **Statistical Modelling in GLIM.** Oxford: Oxford University Press.

Andersen, E.B. (1980) **Discrete Statistical Models with Social Science Applications.** Amsterdam: North Holland.

Anderson, J.A. (1984) "Regression and ordered categorical variables." **Journal of the Royal Statistical Society (B) 46**: 1-30.

Baker, R.J. & Nelder, J.A. (1978) **The GLIM System. Release 3.** Oxford: NAG.

Bishop, Y.M.M., Fienberg, S.E., & Holland, P.W. (1975) **Discrete Multivariate Analysis: Theory and Practice**. Cambridge: MIT Press.

Coleman, J.S. (1964) **Introduction to Mathematical Sociology.** Glencoe: The Free Press.

Cox, D.R. (1970) **Analysis of Binary Data.** London: Methuen.

Dobson, A.J. (1983) **An Introduction to Statistical Modelling.** London: Chapman & Hall.

Duncan, O.D. (1979) "How destination depends on origin in the occupational mobility table." **American Journal of Sociology 84**: 793-803.

Everitt, B.S. (1977) **The Analysis of Contingency Tables.** London: Chapman & Hall.

Everitt, B.S. & Dunn, G. (1983) **Advanced Methods of Data Exploration and Modeling.** London: Heinemann.

Fienberg, S.E. (1977) **The Analysis of Cross-Classified Categorical Data.** Cambridge: MIT Press.

Fingleton, B. (1984) **Models of Category Counts.** Cambridge: Cambridge University Press.

Gilchrist, R. (1981) "Calculations of residuals for all GLIM models." **GLIM Newsletter 4**: 26-27.

Gilchrist, R. (1982a) "GLIM syntax for adjusted residuals." **GLIM Newsletter 6**: 64-65.

Gilchrist, R. (1982b, ed.) **GLIM82.** Berlin: Springer.

Gilchrist, R., Francis, B., & Whittaker, J. (1985, ed.) **Generalized Linear Models.** Berlin: Springer.

Glass, D.V. (1954, ed.) **Social Mobility in Britain.** Glencoe: The Free Press.

Goodman, L.A. (1962) "Statistical methods for analyzing processes of change." **American Journal of Sociology 68**: 57-78.

Goodman, L.A. (1979) "Simple models for the analysis of association in cross-classifications having ordered categories." **Journal of the American Statistical Association 74**: 537-552.

Goodman, L.A. (1981) "Association models and canonical correlation in the analysis of cross-classifications having ordered categories." **Journal of the American Statistical Association 76**: 320-334.

Haberman, S.J. (1974) **The Analysis of Frequency Data.** Chicago: University of Chicago Press.

Haberman, S.J. (1978) **Analysis of Qualitative Data.** Vol. I. **Introductory Topics.** New York: Academic Press.

Haberman, S.J. (1979) **Analysis of Qualitative Data.** Vol. II. **New Developments.** New York: Academic Press.

Healy, M.J.R. (1988) **Glim: An Introduction.** Oxford: Oxford University Press.

Hutchison, D. (1985) "Ordinal variable regression using the McCullagh (proportional odds) model." **GLIM Newsletter 9**: 9-17.

Kroke, D. & Burke, P.J. (1980) **Log-Linear Models.** Beverley Hills: Sage.

Lindsey, J.K. (1973) **Inferences from Sociological Survey Data: A Unified Approach.** Amsterdam: Elsevier.

Lindsey, J.K. (1975) "Likelihood analysis and test for binary data." **Applied Statistics 24**: 1-16.

McCullagh, P. & Nelder, J.A. (1983) **Generalized Linear Models.** London: Chapman & Hall.

Nelder, J.A. (1974) "Log linear models for contingency tables: a generalization of classical least-squares." **Applied Statistics 23**: 323-329.

Nelder, J.A. & Wedderburn, R.W.M. (1972) "Generalized linear models." **Journal of the Royal Statistical Society A 135**: 370-384.

Payne, C.D. (1985) **The GLIM System. Release 3.77.** Oxford: NAG.

Plackett, R.L. (1974) **The Analysis of Categorical Data.** London: Griffin.

Pregibon, D. (1982) "Score tests with applications." in Gilchrist (1982b), pp. 87-97.

Reynolds, H.Y. (1977) **The Analysis of Cross-Classifications.** New York: Free Press.

Robson, D.S. (1959) "A simple method for constructing orthogonal polynomials when the independent variable is unequally spaced." **Biometrics 15**: 197-191.

Upton, G.J.G. (1978) **The Analysis of Cross-Tabulated Data.** New York: Wiley.